Teaching Electrostatics

A Teacher's Resource for Increasing Student Engagement

Donald J. Metz

IDA BOOKS

METZ, DONALD J.
TEACHING ELECTROSTATICS
A TEACHER'S RESOURCE FOR INCREASING STUDENT ENGAGEMENT

Copyright © 2016 by Don Metz
All rights reserved

Illustrations/videos by Don Metz
Photo Credits: © Can Stock Photo Inc.
Cover photo: © Can Stock Photo Inc. / Pukach

ISBN 978-0-9950623-1-3
ISBN 978-0-9950623-0-6 (ebook)

PUBLISHED BY IDA BOOKS
SALMON ARM, B.C.
CANADA

Table of Contents

EI - Essential Information

LA - Learning Activity

TD - Teacher Demonstration

Preface

Introduction to the Teacher	4
How to Use this Resource	5
Curriculum Connections	6

Chapter 1: INTRODUCTION

Introduction to Electricity	7
Careers	8

Chapter 2: WHAT IS ELECTRICITY?

Introduction	10
TD2.1: Attraction of paper bits	11
LA 2.1: Write a Story	12
LA 2.2: Concept Map	14

Chapter 3: THE NATURE OF CHARGE

Introduction	15
LA 3.1: The Nature of Charge	16
TD3.1: Charging an Object	20
EI 3.1: Models in Science	22
EI 3.2: The Electric Atom	26
TD3.2: Charging a balloon	27
EI 3.3: Model of Electric Charge	30

Chapter 4: ELECTROSTATIC PHENOMENA

Introduction	34
TD4.1: Attraction of an Insulator1	35
TD4.2: Attraction of an Insulator2	37
EI 4.1: Insulators and Conductors	38
TD4.3: Separation of Charge	39
EI 4.2: Charging by Contact	40
EI 4.3: Grounding	41
EI 4.4: Charging by Induction	41
LA 4.1: Attraction of Neutral Objects	43
LA 4.2: Build an Electroscope	45
EI 4.5: Testing for Charge	47
LA 4.3: Testing for Charge	47
LA 4.4: Attraction of a Neutral Insulator	48
LA 4.5: Conduction of Charge	49
LA 4.6: Charging by Induction	50
LA 4.7: Inductive Force	54
LA 4.8: Oscillating Paper Bits	56
LA 4.9: Oscillating Foil Bit	57

Chapter 5: THE ELECTROPHORUS

Introduction	59
LA 5.1: Volta's Electrophorus	60
EI 5.1: Charging by Induction	63
LA 5.2: Charging an Electroscope	64
LA 5.3: Oscillating Foil Bit	67
LA 5.4: Lighting an LED bulb	70
LA 5.5: The Leyden Jar	72

Appendix

Equipment	73

Preface

Introduction to the Teacher

This module is intended to promote "minds-on" learning through the active engagement of the learner to link the domain of objects with the domain of ideas. It is a resource that can be utilized across multiple grade levels. Standard outcomes for teaching about electricity in most curricular documents can be found within the objectives. This is a teachers resource with answers. THERE IS A CORRESPONDING SET OF FRAME NOTES FOR STUDENTS. The student frame notes can be downloaded for free from **donmetz.com**. Videos embedded in the document can be found there, as well as on Youtube (see DJM Teach20).

There are three books in this series. In book 1, **Teaching Electrostatics**, the conceptual development of the particle model of electricity underlies an understanding of electrostatics. Students construct their own simple devices, such as an electrophorus, to investigate and explain electrostatic phenomena using the particle model of electricity.

In book 2, ***Teaching Current Electricity,*** a transition from static to current electricity introduces the concepts of electric field, current, voltage, and resistance. Practical, everyday applications such as household electricity complete that module. In book 3, ***Understanding Ohm's Law,*** the historical development of Ohm's law is outlined and the mathematical treatment of current electricity is completed.

How to Use the Resource

In this resource, the following main headings will generally be used: **Introduction**, **Essential Information**, **Vocabulary, Teacher Demonstrations, Teacher notes** (TN/red text), **Learning Activity** and **Did You Know?**

The **Introduction** will set out a context for the concepts to be addressed. **Essential Information** should be recorded, in some manner, in students' workbooks for later assessment. Specialized **Vocabulary** is outlined - for this module the vocabulary is not difficult but must be used very carefully. In the Vocabulary notes, the appropriate usage is explained. The **Teacher Demonstrations** are used to introduce phenomena to students and the **Learning Activity** is a student-centered activity to promote minds-on learning. Teacher demonstrations and student learning activities will be outlined with a rationale, materials for the activity, instructions for the activity, questions to ask, and a summary. **Teacher Notes (TN)**, in red font, provide more detailed explanations, questions, problems, and conceptual development for the teacher's understanding and to incorporate in their instruction as they see fit. All activities are demonstrated in videos and explanations include diagrams and/or animations. Finally, **Did You Know?** presents additional information that may be relevant for some further class discussions or projects.

A student version of this resource that can be used to guide students' learning is also available. The student version includes most of the text in black font that you find in this module and is a set of "frame notes" for students to complete and insert the appropriate diagrams. The student version does not contain the Teacher Notes, this chapter, and the demonstration videos (students should do these activities themselves).

Curriculum Connections

A set of general objectives are provided at the beginning of each chapter that can be linked to various grade levels depending on local curriculum. Teachers may modify or omit objectives as they deem appropriate.

The objectives followed in this module were developed using many of the Next Generation practices essential for learning science. These include: developing and using models, planning and carrying out investigations, analyzing and interpreting data and constructing explanations (see A Framework for K-12 Science Education: Practices, Crosscutting Concepts, and Core Ideas, Committee on Conceptual Framework for the New K-12 Science Education Standards; National Research Council, page 42 (available from The National Academies Press at http://www.nap.edu/catalog.php?record_id=13165).

Introduction

Introduction to Electricity

Imagine living in a world without electricity. No television, no lights, computer or smartphone, and no texting! Electricity is a fundamental asset in virtually all aspects of our life. It provides light and heat, and is the basis for all of our communications. Importantly, understanding electricity is an essential component for many careers such as electricians, engineers, and transmission line personnel.

Careers - ELECTRICITY

Electrical energy powers our homes and businesses supplying energy to everything from light bulbs and refrigerators to manufacturing plants and farm operations. With such a wide variety of applications, employment opportunities in the electrical sector are many and diverse. We find regulated professionals such as electrical engineers in addition to apprentice trades for electricians and auto mechanics.

Some professions require a university or college education and some require on-the-job training, and certainly all require some high school background. Beginning to understand electricity opens a gateway to a wide range of opportunities. Indeed, the pending retirement of the "baby boom" generation will open many doors for young skilled new workers. Now is a good time to begin to develop an understanding of electricity.

Great opportunities exist

CHAPTER 2

What is Electricity?

Introduction

WHAT IS ELECTRICITY? Have you ever thought of where it comes from? Electricity seems to be all around us and influences our everyday experiences on a regular basis. We are not the first to ask "What is electricity"? The ancients, who knew about the natural forms of electricity such as lightning, began to think about the nature of electricity. Over two thousand years ago, Greek philosophers discovered that when amber was rubbed by certain materials, small bits of straw were attracted to the amber. Immediately, they began to think about how to explain this curious phenomenon.

TN: Most people will use the term electricity as a "catch all" term without understanding the nature of electricity. In this chapter students are encouraged to speculate about "what is electricity" and to describe their own personal experiences.

Objectives

1. Introduce students to the phenomena of electrostatics.

2. Students reflect on their experiences and ideas about electricity.

TEACHER DEMONSTRATION 2.1: Attraction of small bits

Materials: PVC Plastic rod, cloth, paper bits.

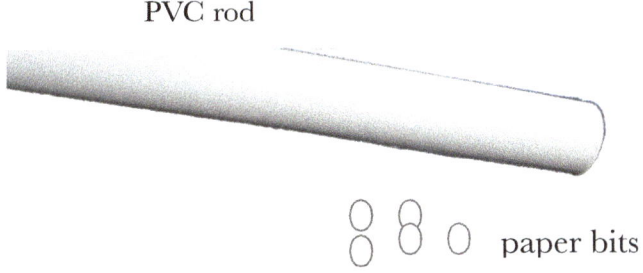

TN: PVC plastic can be obtained inexpensively at the local hardware store. It can be easily cut to length. A wool, or wool like cloth, is best and paper bits are easily found in photocopier machines or in hole punch devices.

Instructions for the Activity

1. Rub the PVC plastic rod with a wool cloth and bring it near the paper bits.

TN: The paper bits will be attracted to the rod. It is quite possible that the bits will oscillate between the rod and the table. The explanation for this action will be addressed later.

Question:

TD2.1a.   Why do you think the paper bits are attracted to the plastic rod?

TN: At this time students should be encouraged to speculate about the nature of the forces of attraction. Some might say "it's electricity" without knowing what electricity really means. It is not necessary at this time for students to articulate a formal explanation of electricity.

LEARNING ACTIVITY 2.1: Write a Story

TN: A student once told me a story about a group of children who were holding hands while folk dancing in the gym. In order to protect the gym floor each student was required to wear felt boots. As they were dancing one of the children reached out and touched the metal volleyball standard with their toe. A giant spark jumped from their foot to the floor and it knocked the entire group over!

What's your story?

Materials: Your imagination and a place to write.

Instructions for the Activity

1. Each student writes a short story about a personal experience with electricity.

2. Share your story with your classmates and family.

Questions about your story

LA2.1a.   Can you explain what happened in terms of electricity?

LA2.1b.   Was it a dangerous situation?

LA2.1c.   What can be done to prevent the situation from happening again?

TN: You may be surprised at the experiences students have had with electricity. This exercise provides an excellent means to discuss electrical safety early with your students.

All of the students in your class have likely experienced, in one way or another, the phenomena of electricity. Telling a personal story to describe their experiences is an excellent way to generate interest and integrate other disciplines, such as language arts, with science. Many students will tell of experiences with family members who shuffle across a rug and touch a doorknob to experience a shock. Some might tell of sticking things in electric sockets (NO! .. Dangerous!), or touching their tongue to the terminals of a 9v battery (Again .. NO!).

Students should be CAUTIONED at this time that electricity can be dangerous and that many people die every year as a result of electrocution. According to the Electrical Safety Foundation International organization more than 30,000 non-fatal shock accidents occur each year and every day many children are treated in hospital emergency rooms for electrical shock or burn injuries caused by electrical incidents. Many more suffer from fires caused by electrical malfunctions. Now is a good time to make sure students recognize the warning signs of an electrical hazard. Visit the Electrical Safety Foundation International (ESFi.org) for more detail on home and workplace safety.

LEARNING ACTIVITY 2.2: Concept Map

Make a concept map to help organize your ideas about electricity. Include some aspects of your story in your map.

Materials: Paper and pen

Instructions for the Activity

1. Circle your idea and connect ideas together with lines Add as many words, boxes, and arrows as you can.

TN: Concept maps (Mind maps) are useful graphical organizers of knowledge. Concept maps depict ideas and the links between related ideas. Students can add their ideas as they progress through the activities in this module for reflection and review. Post student concept maps on the board. Students can also work in groups and prepare large concept maps on newsprint.

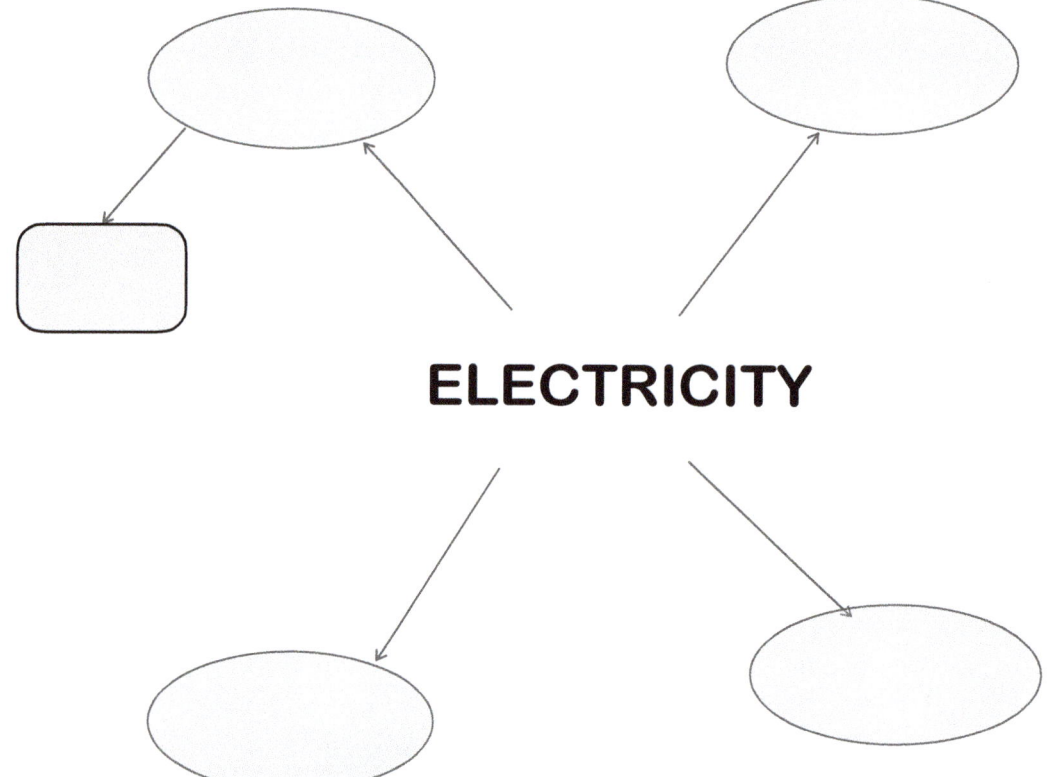

CHAPTER 3

The Nature of Charge

Introduction

Have you every pulled your sweater over your head and made your hair stand on end? Have you ever rubbed a ballon on your hair? Have you heard people use the term electrical charge? Do you wonder what is charge and how we know it exists? In science we should always ask ourselves, "What are the good reasons to believe?" and "how do we know?" In the following activities we will try to demonstrate the evidence for the existence of electric charge.

Objectives

1. Provide evidence for the existence of electric charge.
2. Discuss early models of electricity, eg. Plutrach's model, Gilbert's model, Franklin's one-fluid model, Dufay's two-fluid model, and the particle model.
3. Relate the particle model of electricity to atomic structure.

Vocabulary

In this activity you must be very careful to distinguish between bringing objects "nearby" and "touching".

<u>Nearby</u> - is approaching an object <u>slowly</u> but never touching it.

<u>Touching</u> - is making contact with an object.

TN: It is good practice to have students' draw diagrams to illustrate their observations and explanations. In these diagrams, every time an object changes from being nearby to touching, or touching to nearby, a new diagram is required - <u>this is important!</u> In this teacher's resource, teacher diagram examples will be illustrated for each activity.

Learning Activity 3.1: The Nature of Charge

TN: This activity helps introduce students to electrostatics by charging common materials. Most textbooks merely claim there are two types of charge and provide no evidence for the existence of only two types of electric charge. In this activity, students investigate the effects of electric charge and speculate on the existence of a third type of charge.

Students could be assigned this activity as a home experiment or as a classroom activity. Hand out the pieces of scotch tape (tape them to the students' desk) while the students read their instructions. Demonstrate how to fold the tab and charge the tapes (see video). Ask students to answer the questions carefully. After some exploration time for the students, provide a demonstration and answer each question with the class.

TN: Using models is an essential part of doing science. Scientists (and students!) invent models, and then test them to determine if the observations and data match the prediction of the model. If the observations and data match the model's prediction then we have confidence that our model is a good explanation of the phenomena under investigation. We say that our observations and data become evidence. If the evidence does not support the model then we have a discrepant event, and we look to modify or completely change our model.

Materials: Scotch tape, a partner (fellow student, Mom or Dad can help too!)

Instructions for the Activity

1. Take a piece of scotch tape and fold 0.5 cm at one end of the tape to form a tab. Make a second piece the same way.
2. Tape the two pieces, one on top of the other to a smooth surface and mark the top one T and the bottom one B. (figure 3.1)
3. Peel them off together quickly.
4. Separate the pieces of tape from each other.

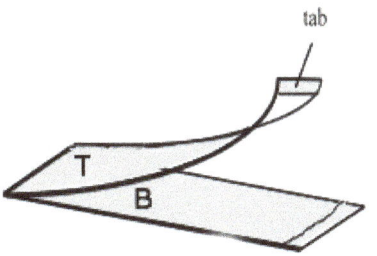

Figure 3.1: Scotch tape.

TN: This is a good activity for a homework assignment.

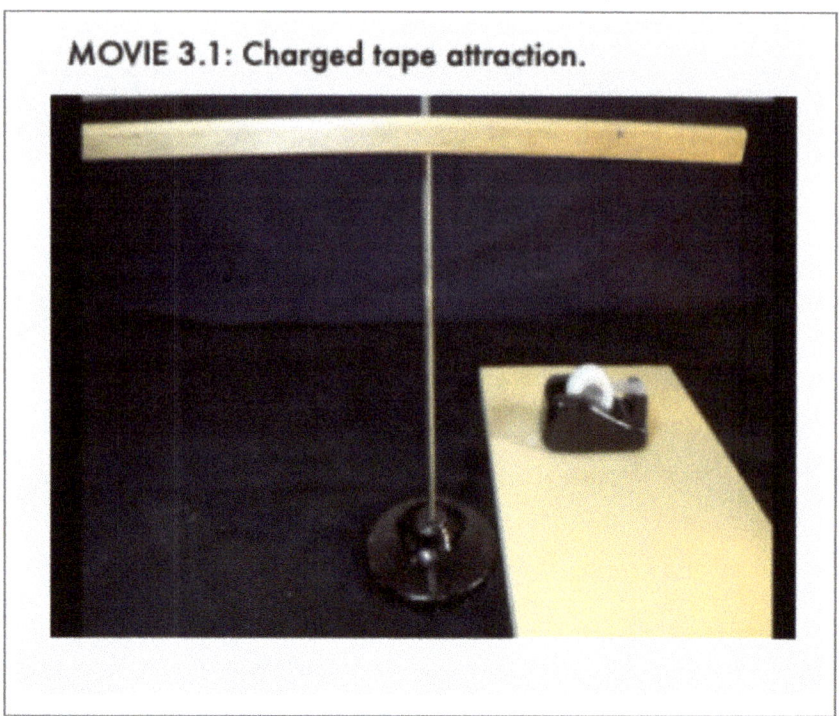

https://youtu.be/i7Lu8h-1kAg

Questions

<u>LA3.1a.</u> Bring the pieces of tape nearby each other SLOWLY. What happens?

TN: The Top and Bottom tapes attract each other. Note: you can do this with a partner or you can suspend the tapes from a support as shown in the video.

Continue your investigation...

5. Prepare another identical pair of tapes for your partner (or place them on the support.

Questions

<u>LA3.1b.</u> Bring your Top tape slowly towards the Top tape of your partner and record your observations.

TN: The Top tapes repel each other.

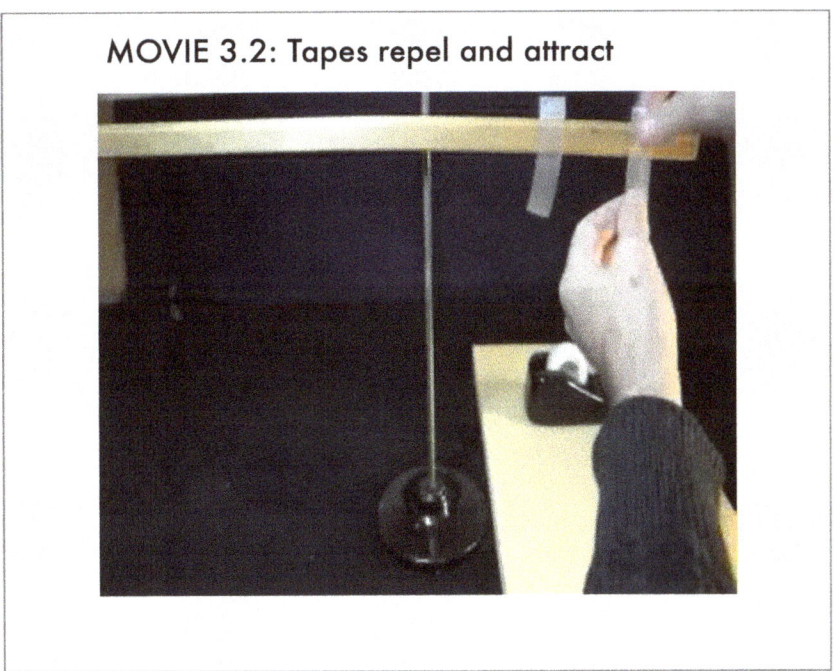

https://youtu.be/r5xDXe62JSESummary

<u>LA3.1c.</u>   Bring your Bottom tape slowly towards the Bottom tape of your partner and record your observations.

TN: The Bottom tapes repel each other.

<u>LA3.1d.</u>   Bring each Top tape nearby each Bottom tape. What happens?

TN: The Top and Bottom tapes attract each other.

<u>LA3.1e.</u>   Discuss with your partner why the tapes moved. Record your ideas for class discussion. Summarize your results and make a rule to predict the force on the tapes.

TN: There are two kinds of charge, Top and Bottom. Some students will use the terms positive and negative right away but it is not necessary at this point. Call them Fred and Barney if you wish! The top and bottom tapes attract each other. The top tapes repel each other and the bottom tapes repel each other. No other effects are observed.

Questions

<u>LA3.1f.</u>   If a third type of charge existed, what would it do to the top and bottom tapes?

TN: If a third type of charge existed it would repel both the top and bottom tape. We never observe this so we conclude that only two types of charge exist.

Teacher Demonstration 3.1: Charging an object

<u>Materials:</u> plastic rod, silk like material, pvc pipe, wool like material, electroscope (if you do not have an electroscope see "Building an Electroscope" in the next chapter.

TN: This activity demonstrates charging, like and unlike charges, attraction and repulsion. It is very similar to the tape activity but using different materials.

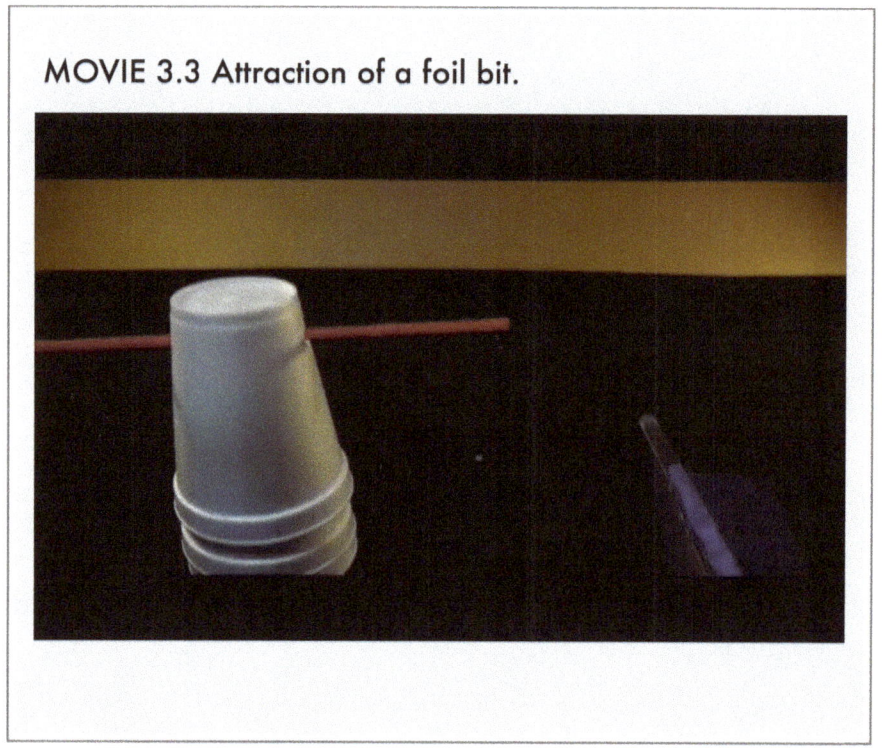

MOVIE 3.3 Attraction of a foil bit.

https://youtu.be/MugLW9MoDGE

Instructions for the Activity

1. Rub the plastic rod with a silk like material to put a charge on the rod.

2. Bring the charged rod nearby and touch the foil bit (or pith ball) on the electroscope.

TN: When we bring the charged rod nearby the foil bit it is initially attracted and then touches the rod. When the foil bit touches the rod it becomes charged the same as the rod and the foil bit and rod repel each other.

Continue ..

3. Rub the PVC rod with a wool like material and bring it nearby (but do not touch) the charged foil bit.

TN: When we bring the PVC pipe nearby it is attracted to the foil bit. Charge is the property that gives rise to the electrical forces of attraction and repulsion. We say that the plastic rod is positively charged and the PVC pipe is negatively charged (the names positive and negative are historical). The scotch tape and the charged rods both demonstrate that like charges repel and unlike charges attract. Note that we have only demonstrated the phenomenon and we have not provided an explanation of the underlying mechanism. Ask students now to speculate on the question "what is charge?"

TN: If you do not have an electroscope see instructions on building an electroscope in Movie 4.5 in the next chapter. At this time, this activity is a teacher demonstration. Students will build their own electroscopes in chapter 4.

Essential Information 3.1: Models in Science

A scientific model is a representation which stands for, and helps explain, other things. A model can be physical (a real thing), imagined (in my brain!), or mathematical. In science, we develop models which have explanatory and predictive powers (like the model of the universe or the atomic model of matter) and we test these models in the world around us. If our model predicts our observations we accept the model as a valid description of our world. However, if our model encounters discrepant events and fails to provide adequate explanations we begin to modify our model or search for an entirely different model.

Observations are used to test models, either in the world around us or by thought experiments as we re-think and apply our model to new, and sometimes discrepant situations. Our observations lead us to identify regularities and patterns in nature. A good model explains a lot of things and survives the test of time. If you use a model based approach to science you are beginning to think like a scientist.

Early Models of Electricity

Electric phenomena can be explained in terms of many models.

A. The Greek philosopher Plutarch (A.D. 100) believed the attraction of small bits of straw to rubbed amber was the result of heated air that swirled around and pushed the straw to the amber.

Questions

<u>EI 3.1a.</u> Rub your hands together rapidly. What do you feel?

TN: When you rub a material with another material heat is created by friction. Plutarch's argument, in his time, had a measure of plausibility.

<u>EI 3.2b.</u>   Draw Plutarch's model and suggest an experiment you could do to test Plutarch's model.

TN: Students typically say do the test in a vacuum. Inform the students that these are very good scientific ideas however the Greeks could not make a vacuum. Other possibilities include do the test in the cold, or do the test under water. All are good thoughtful answers and students who provide these suggestions should be recognized for practicing high grade scientific thinking.

<u>Plutarch's Model</u>

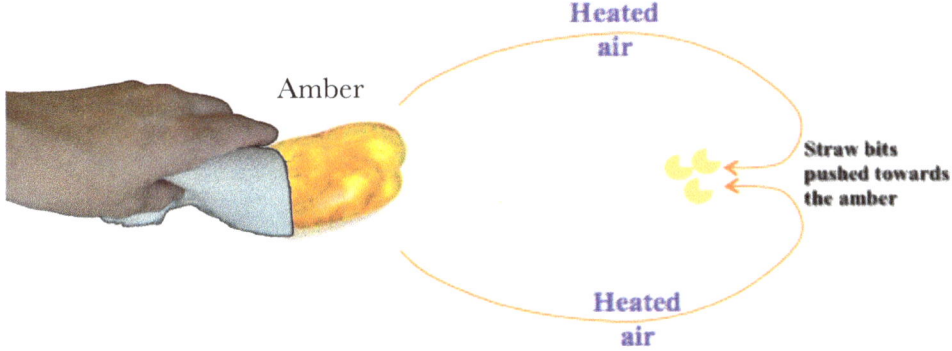

B.   John Gilbert, a British philosopher (1600) suggested that the rubbed amber gave off an "effluvium" which emanated from the amber to nearby objects. The effluvium stuck to the object and pulled it back.

Questions

<u>EI 3.1c.</u>   Draw Gilbert's model and suggest an experiment you could do to test this model.

TN: Students may suggest that you could put something between the rubbed amber and the nearby object. Again, this is an excellent idea for testing the model.

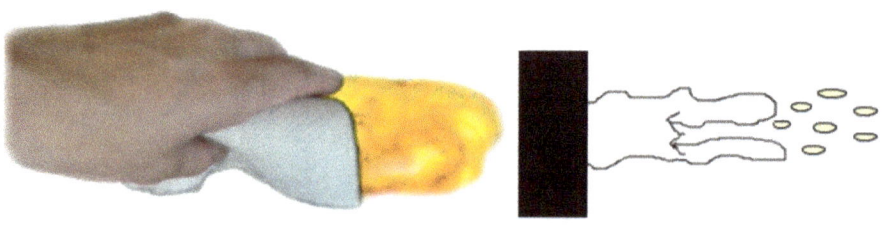

C. <u>Ben Franklin's One Fluid Model</u>

Ben Franklin proposed that a neutral object has a balance of electrical fluid. If an object acquires excess fluid it is positively charged and if an object has a deficit of fluid it is negatively charged.

Questions

<u>EI 3.1d.</u>   Can you explain the results of the scotch tape experiment using Ben Franklin's one fluid model?

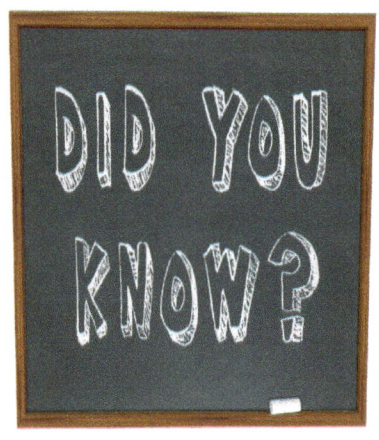

TN: It is a useful exercise in critical thinking to challenge students to explain the results of the scotch tape experiment in terms of an electrical fluid. A good answer would be – when I press down on the top tape I am pushing electrical fluid into the bottom tape. Therefore, the top tape has less fluid and is negatively charged and the bottom tape has more fluid and is positively charged. In terms of a model of electric charge, Franklin's one fluid theory explained almost all of our electrostatic observations and introduced the fundamental idea of conservation of charge.

In 1752 Benjamin Franklin is alleged to have conducted his famous kite experiment in Philadelphia, USA. Franklin's experiment was extremely dangerous and other people who tried the experiment were electrocuted. The popular TV show Mythbusters demonstrates the dangers of lightning strikes in a mock up of Franklin's experiment (Mythbusters, season 4 episode 9, Franklin's Kite). Franklin recognized the dangers of lightning and subsequently invented the lightning rod to divert lightning strikes to the ground.

D. <u>Dufay's two fluid model</u> – about the same time as Franklin, Charles DuFay, a French scientist, postulated that there are two kinds of electricity, vitreous (produced by rubbing glass) and resinous (produced by rubbing a resin like amber). He stated that neutral objects have equal amounts of fluid and if an object has more of one fluid than another it becomes charged.

E. <u>The Particle Model</u> – similar to the two fluid model, the particle model of electricity states that there are two kinds of particles (known today as positive and negative). If an object has more of one type of particle than another it becomes charged.

TN: Franklin's one fluid model and the two fluid model of electricity explain most electrostatic phenomenon quite well. They were excellent models in their time. Today, because of important discoveries (like Thomson's discovery of the electron, Millikan's oil drop experiment, and Rutherford's gold foil experiment) and the development of the atomic model of matter in the early part of the 20th century we have come to accept a particle model of charge. Robert Millikan, using charged oil drops, was able to demonstrate the discrete nature of electric charge. That is, the smallest amount of charge that you can add or take away is always a single particle. J.J. Thomson's experiment, using cathode rays, showed that this particle was the electron. Ernest Rutherford, fired positively charged alpha particles at a gold foil to demonstrate that the nucleus of the atoms in the foil was a stable positively charged mass that deflected the alpha particles.

Robert Millikan   J.J. Thomson   Ernest Rutherford

Essential Information 3.2: The Electric Atom

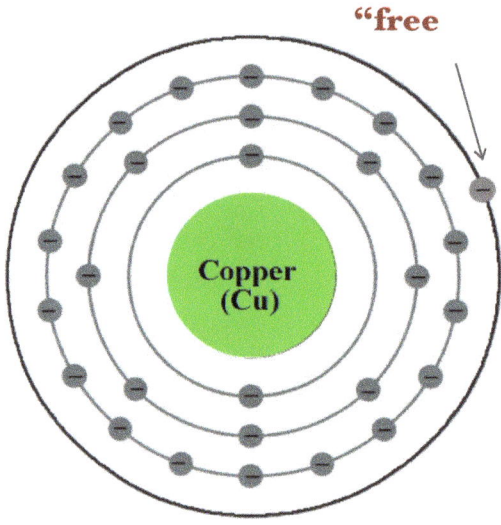

The Bohr model of the atom is a useful model to think of matter today. There is a positively charged central nucleus of protons and neutrons surrounded by orbital shells of negatively charged electrons. Many atoms, like the copper atom illustrated above, have a "free" electron in the outer orbit. This electron moves very easily from atom to atom and by using a separation of charges we can make these electrons move. If we have extra electrons an object will be charged negatively, if we lose electrons an object will be positively charged.

TN: Everything depends on the movement of charges. Continually, and repeatedly, ask your students "where do the negative charges go?"

Electrostatics Demonstrations

All materials are made of atoms and therefore have positive and negative charges. The number of charges in any object or material is huge - billions upon billions. For example in the human body we have approximately 7×10^{27} atoms, that is seven billion billion billion!!

However it is the balance of charge that is important. If the number of positive charges equals the number of negative charges the electric effects cancel each other and the material is neutral.

Teacher Demonstration 3.2: Charging a balloon

Materials: balloon, hair

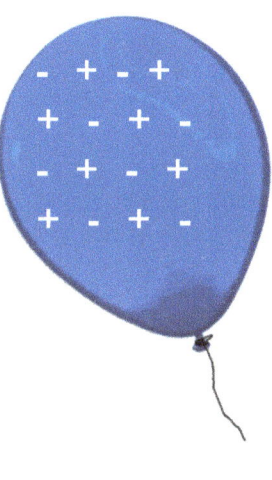

TN: Normally our hair would have equal positive and negative charges and be neutral. Likewise for a balloon (and most other materials).

Instructions for the Activity

Rub a balloon vigorously on your hair. Negative charges (electrons) move from your hair to the balloon. The balloon now has more negative charges than positive charges and becomes negatively charged. The hair loses negative charges and becomes more positive (individual strands of hair may even repel each other).

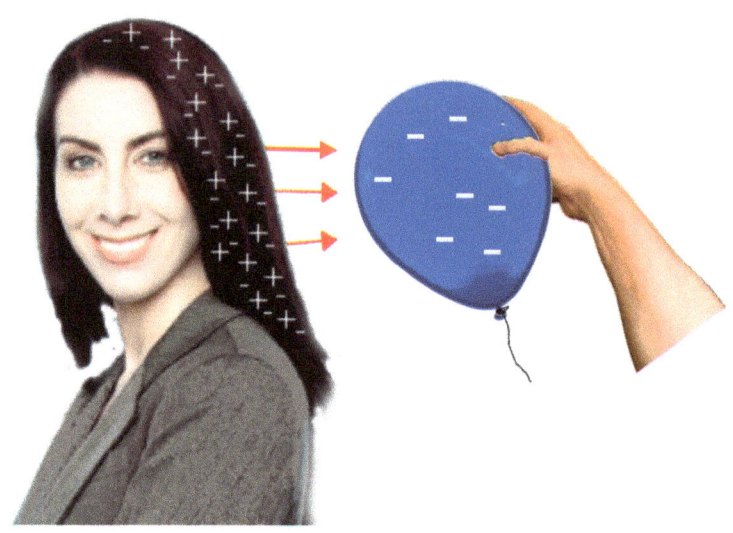

Give it a try!

Bring the balloon towards a wall. The wall is initially neutral with equal numbers of positive and negative charges.

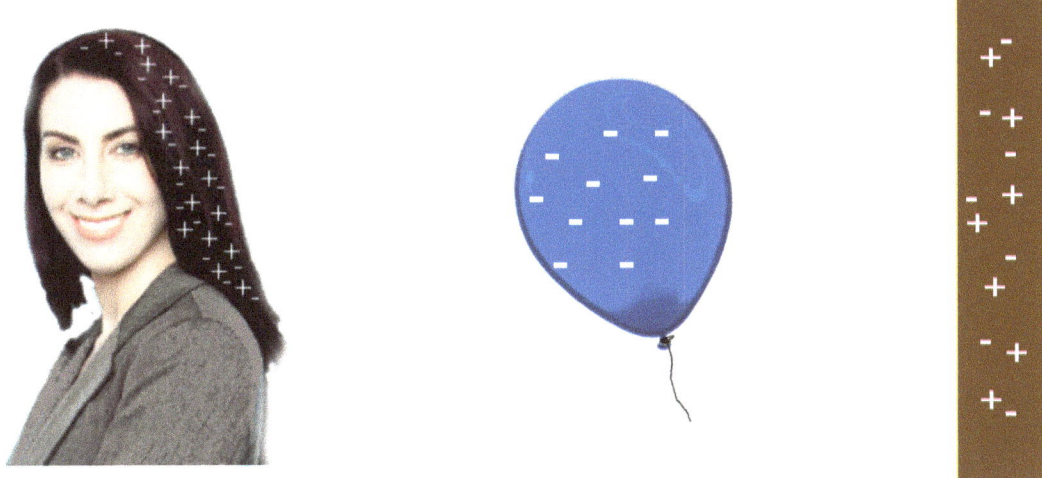

As the balloon approaches the wall the negative charges in the wall move further away (just a bit) repelled by the negative charges on the balloon. Every negative charge leaves behind a positive charge near the edge of the wall. The negative balloon and the positive surface of the wall attract each other and the balloon will stick to the wall.

Essential Information 3.3: Model of electric charge

Before we continue, let's summarize what we know about the model of electric charge so far.

1. There are two kinds of charge, <u>positive and negative.</u>

TN: The assignment of the names positive and negative is purely arbitrary. Historically, glass rubbed with silk was called positive and ebony rubbed with fur was called negative. Anything that repelled the charged glass would be positive and anything that repelled the charged ebony would be negative. Ben Franklin was the first to use these terms.

2. Charge is <u>conserved,</u> it cannot be created nor destroyed.

TN: A very common misconception is that charge is used up. Charge is **always** conserved and can only be displaced from one place to another. The energy of moving charges is what gives rise to heat and light.

3. A neutral body has an <u>equal</u> number of positive and negative charges.

TN: Neutral bodies have equal numbers of positive and negative charges but any object has **millions and millions** of charges within the atoms of the material. For convenience we only show a few positives and an equal number of negatives to indicate the object is neutral. At times, textbooks will even represent neutral charges without any illustration of charge. In the diagram on the right when no charges are shown the object is said to be neutral – however even though no charges are illustrated there are still LOTS of charges.

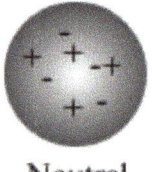
Neutral

Neutral

4. In solids, positive charges are <u>fixed</u> and negative charges are <u>free</u> to move.

TN: Positive charges can actually move in ion solutions. However, in this unit we are only dealing with solid materials.

5. An <u>excess</u> of negative charges produces a negatively charged body. A deficit of negative <u>charges</u> leaves a positively charged object.

TN: Emphasize that only negative charges move. To become negatively charged an object must receive negative charges and to become positively charged an object must give up negative charges. In diagrams, sometimes only the net charge is illustrated. Remember to emphasize there are LOTS of both kinds of charges on any charged object and the negative/positive signs only indicates more negatives or more positives.

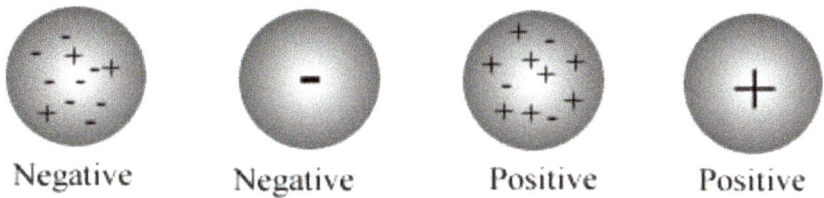

6. Charge is shared by <u>contact</u>. Materials, which allow charges to move easily, are called <u>conductors</u>. Materials in which charges do not move easily are called <u>insulators.</u>

TN: Typically, we think of conductors as metals such as copper, and insulators as non-metals like plastic. However, remember that many materials are not either one or the other, it is a matter of degree.

7. Like charges repel and unlike charges attract.

In diagrams, we usually indicate the forces of attraction and repulsion with arrows in the direction of the force. Remember that the charges act on each other.

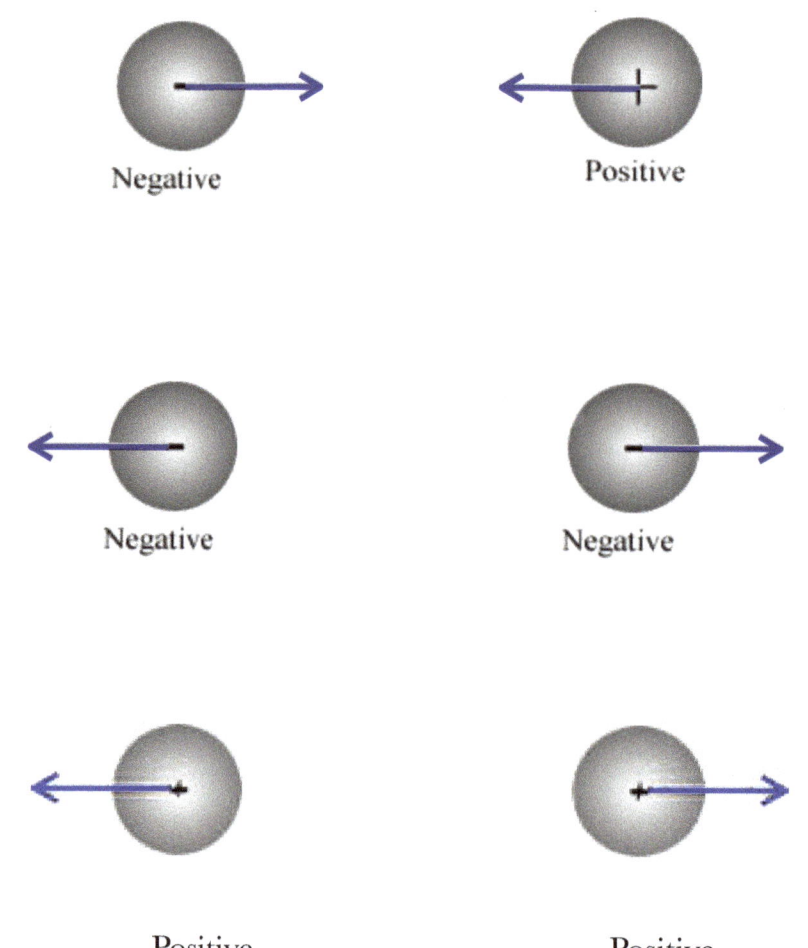

TN: The charges always act with an equal and opposite force.

TN: Now is a good time for a quiz. Students should know these seven fundamental principles of the model of electric charge before doing their own experiments in the next chapter. The teacher can easily cover this chapter in a typical class to prepare students for the fun activities to come.

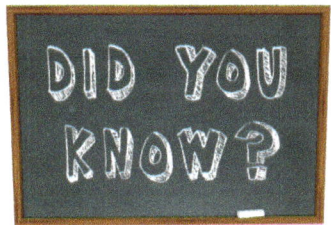

Triboelectric Series

There are many kinds of materials that can be charged by rubbing them together. One material loses negative charges (electrons) and becomes more positive while the other material gains negative charges and becomes more negative. The capacity of a material to gain or lose electrons is dependent on a number of factors. For example, some plastics may become positively charged by friction while other kinds of plastic become negatively charged. The dye used in certain cloths can also affect their ability to lose or gain electrons. Wood from different trees may also have different properties. Human skin sometimes becomes more or less oily with hand cream or other cosmetics affecting the ability to charge an object. The capacity of a material to acquire or lose electrons is called the triboelectric series. Here is a sample series with the more positive materials near the top. A triboelectric series can vary considerably, the best way to choose materials to work with is to try them.

More +
hair
glass
synthetic cotton
silk
paper porcelain
wood
cork
acrylic cloth
Styrofoam
plastic bag
drinking plastic straw
rigid acrylic
PVC tube
hard rubber
More -

CHAPTER 4

Electrostatic Phenomena

Introduction

There are two kinds of electric particles, positive and negative. In solid materials, only the negative particles can move. We can explain most electrical phenomena by thinking about the movement of negative charges. When negative charges move from one place to another, one region becomes more negative and the other region becomes more positive.

Objectives

1. Demonstrate and explain electrostatic phenomena using the particle model of electricity. Include: conservation of charge, conduction, grounding, attraction of a neutral insulator, induction.

TN: Students now have a simple particle model of electricity that they can use to explain electrical phenomena. It is best to begin with a few basic teacher demonstrations and summarize the particle model of electricity to help students explain some of the fundamentals. Then, students should complete the electrostatics hands-on activities that follow.

The teacher performs the demonstration and asks students to explain in terms of the particle model of electricity. By introducing an electric force (that is, by bringing a charge nearby), negative charges can be displaced and one region becomes more negative while another region becomes more positive. The resulting net charges exert electrical forces usually causing something to move. Repeatedly ask the question, "Where do the negative charges go"?

Teacher Demonstration 4.1: Attraction of a Neutral Insulator

Materials: Charged object, stream of water

Instructions For the Activity

1. Bring a charged object (like a balloon rubbed on your hair or a charged rod) near a fine stream of water.

Question

What do you observe? Can you explain using the model of electric charge?

TN: The water stream is attracted to the balloon. The water molecules are polar and the charges can align themselves such that the positives are closer to the negative rod resulting in a net force of attraction.

TN: We already know from our everyday experiences that water is electrically neutral and that a balloon rubbed on our hair will become charged. As simple as this demonstration is, students find it very dramatic.

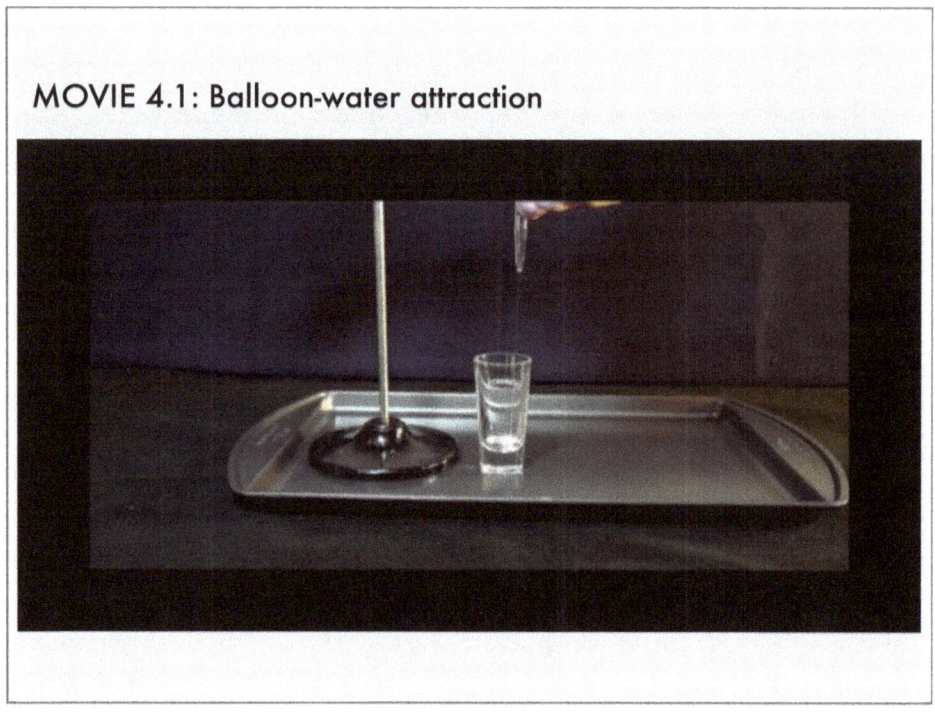

https://www.youtube.com/watch?v=slWDXjOSDts

Diagram

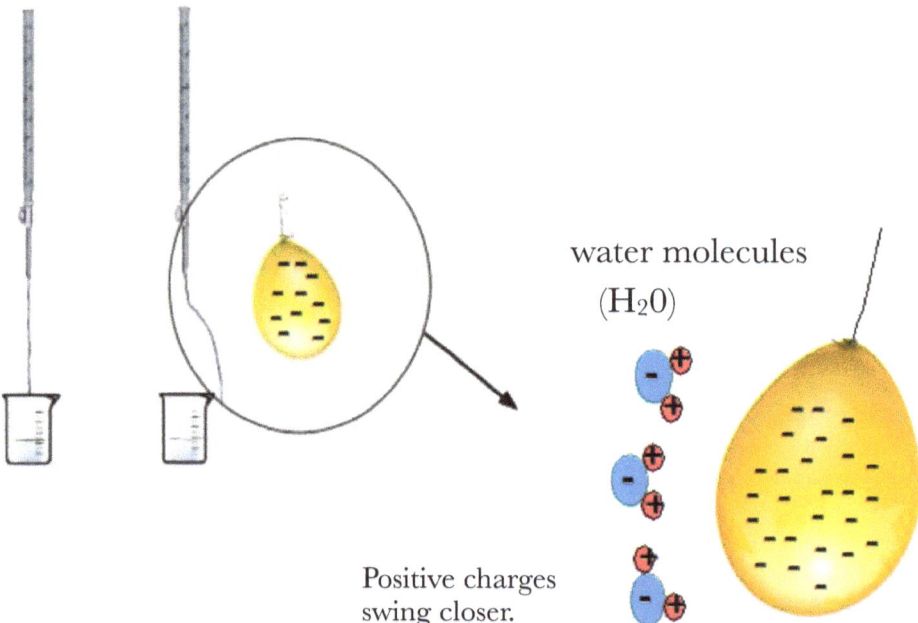

Teacher Demonstration 4.2: Attraction of a Neutral Insulator

Materials: charged rod, 2" x 4" x 8' piece of wood (or similar), bottle cap

Instructions for the Activity

1. Support a 2" x 4" (8 feet long!) on a watch glass or on a plastic bottle cap (place the plastic bottle cap upside down on the table such that it spins easily).
2. Bring a charged object nearby to make the board turn.

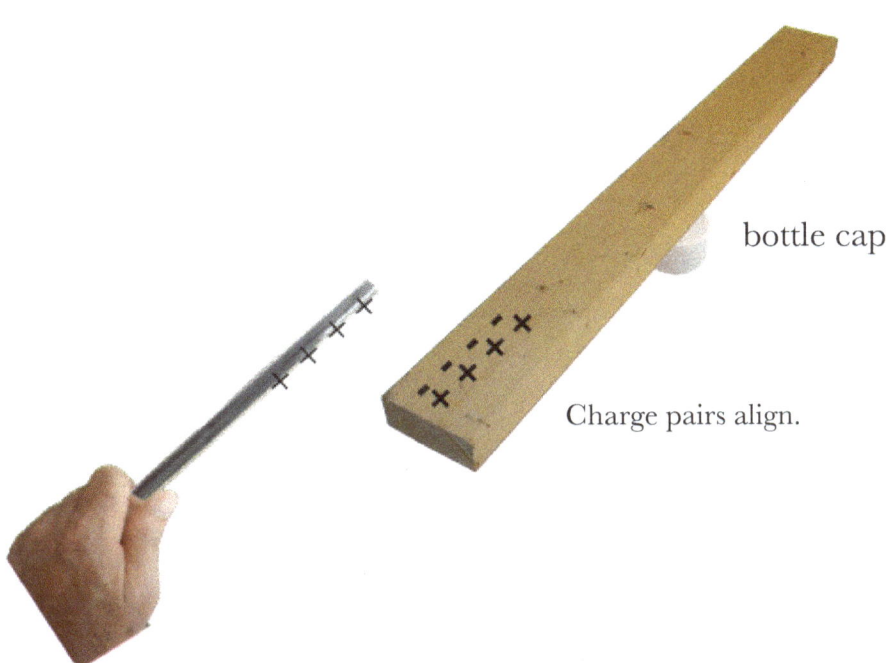

TN: This demonstration is very easy to do and dramatic. Even though wood is an insulator the charges can still move around enough to create a net force of attraction.

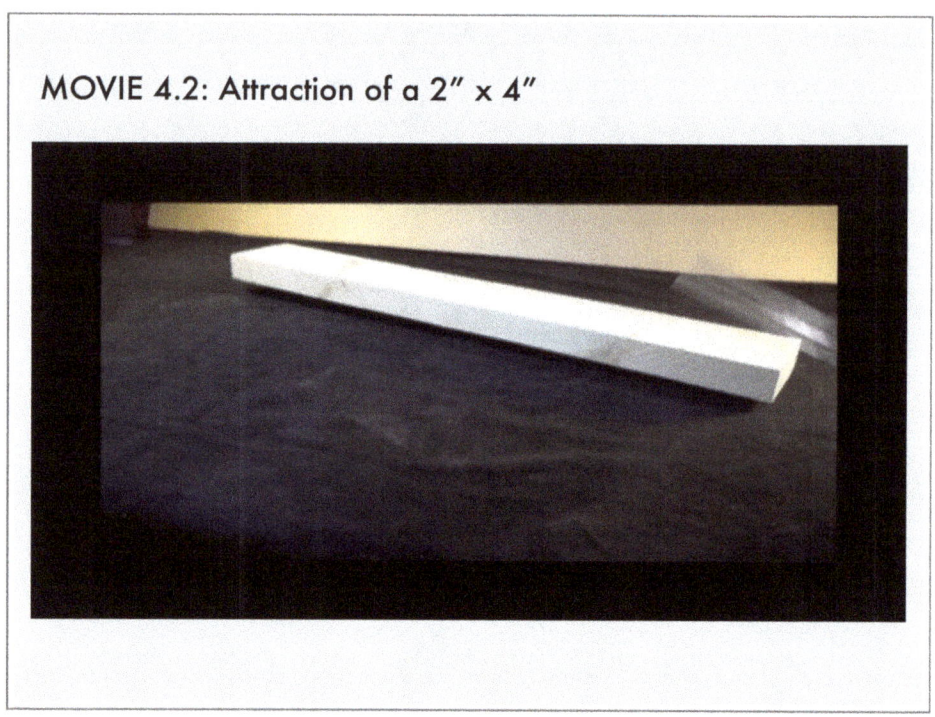

https://www.youtube.com/watch?v=5SwMXU4lups

Essential Information 4.1: Attraction of Insulators and Conductors

On insulators, positive charges are fixed and negative charges do not move very much. However the positive/negative pair will align and still result in a net force of attraction. That is, a charged object can attract a neutral object (just like the water stream or the 2 x 4 demonstration). We say that the charges are polarized.

On solid conductors, positive charges are fixed and negative charges move freely. The negative charges will physically separate from the positive charges. Again, the result is a charged object can attract a neutral object but the effect is much stronger for a conductor.

TN: Materials are not either conductors or insulators but form a continuum from good conductors (like copper) to poor conductors, and poor insulators to good insulators (like glass).

Teacher Demonstration 4.3: Separation of charge (conductor)

Materials: charged rod, foil bit (see instructions for building an electroscope)

Instructions for the Activity

1. Bring a charged rod nearby a foil bit

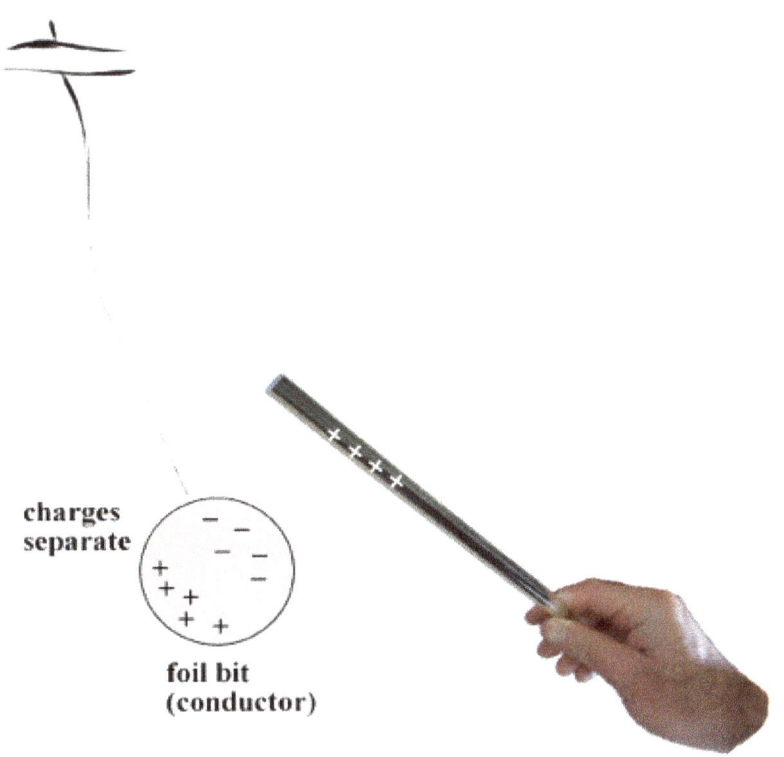

TN: You can easily demonstrate the attraction of a conductor by bringing a charged rod nearby a foil bit. It is very common for the foil bit to be attracted to the rod quickly, touch the rod, and then repel. At this time, the foil bit has become charged (see following diagrams). You can neutralize the foil bit any touching it. Show students how to slowly bring the charged rod close by without the rod and foil bit touching each other.. The construction of a foil bit and the electroscope used in the next activities is described in Learning Activity 4.2 and MOVIE 4.5. The teacher will need one for these demonstrations and the students will build theirs later.

Essential Information 4.2: Charging by Contact

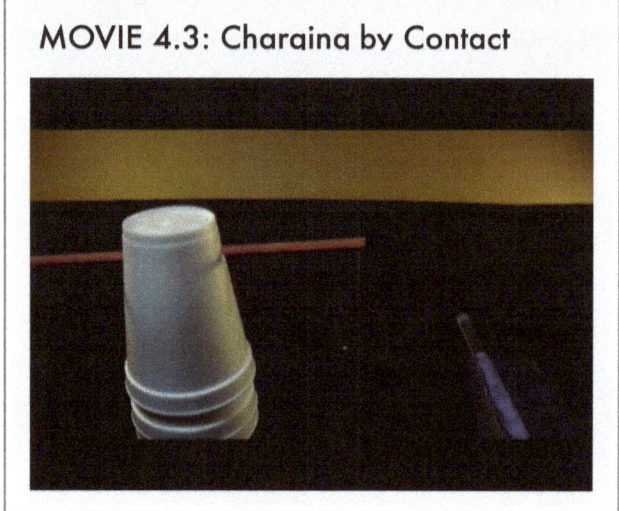

MOVIE 4.3: Charging by Contact

If a charged object touches a conductor there is a sharing of charge between the object and the conductor.

DIAGRAM YOUR EXPLANATION (every time a change occurs from touching to nearby, or nearby to touching, make a new diagram.

https://youtu.be/OUGVSrae7D8

TN: Charging by contact diagrams.

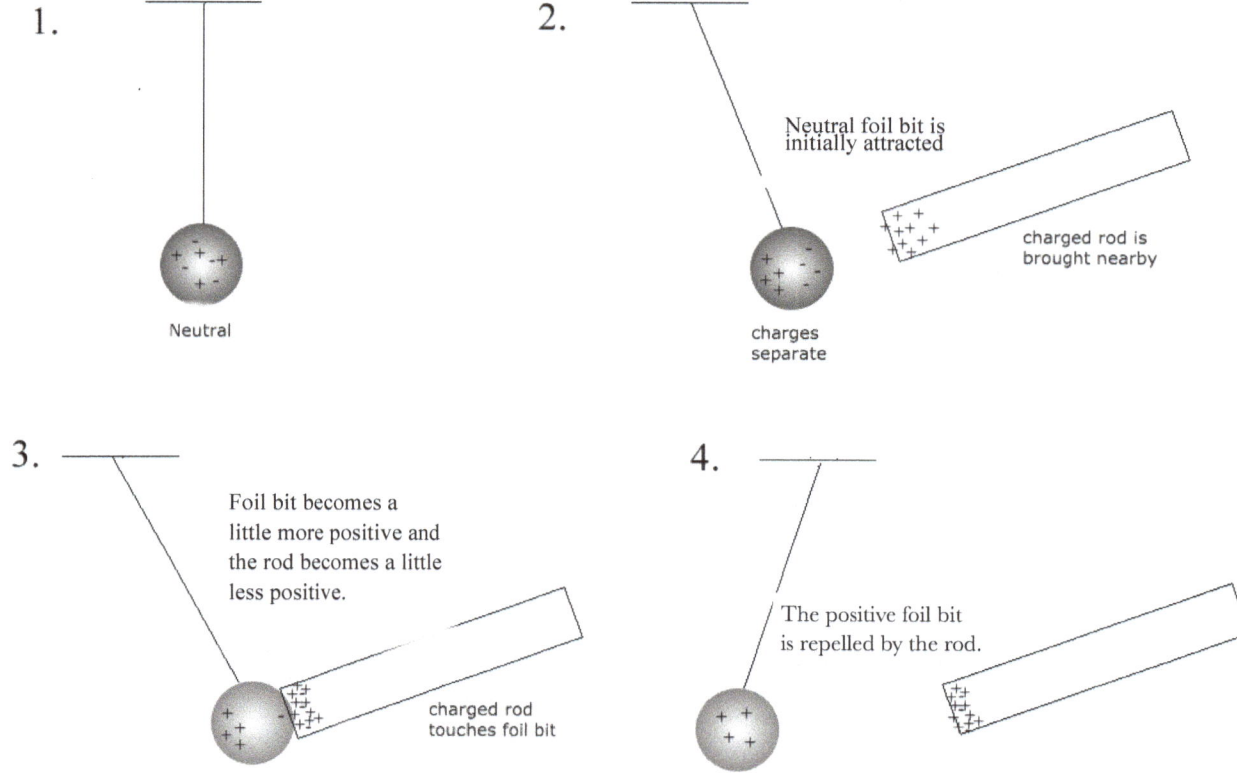

1. Neutral

2. Neutral foil bit is initially attracted / charged rod is brought nearby / charges separate

3. Foil bit becomes a little more positive and the rod becomes a little less positive. / charged rod touches foil bit / Some negative charges are transferred to the rod.

4. The positive foil bit is repelled by the rod.

Essential Information 4.3: Grounding

Grounding is the sharing of charge with an object. Often the earth is used as a ground. In the following examples, because of the small amount of charge, your finger will do. However, NEVER use your finger to ground an electrical object in any other cases - especially household circuits (For more information about home and workplace safety visit the website www.ESFi.org). Excess charges flow from, or to the ground, neutralizing the grounded object. The earth is such a large object that if it loses or gains some charges it remains essentially neutral. A ground can also be used as a reservoir of charge.

Clearly distinguish between objects that are nearby and objects that are touching. Use the terms conductor and insulator, and separation and polarization (aligning of charge).

TN: VOCABULARY

It is essential to carefully use and emphasize correct vocabulary. Students should clearly distinguish between objects that are nearby and objects that are touching. They should use the terms conductor and insulator, and separation and polarization (aligning of charge). Graphic organizers such as mind maps and concept frames are useful to emphasize vocabulary.

Note: Remember that the distinction between a conductor and an insulator is not either/or but a distinction that is a matter of degree. There are materials that are good conductors and materials that are poor conductors as well as materials that are good insulators and materials that are poor insulators.

Essential Information 4.4: Charging by Induction

If a charged object is brought nearby a conductor the charges on the conductor will separate. If the conductor is then grounded with the charged object nearby, negative charges will flow from the conductor to the ground, or from the ground to the conductor, depending on the charge nearby.

TN: Grounding diagram

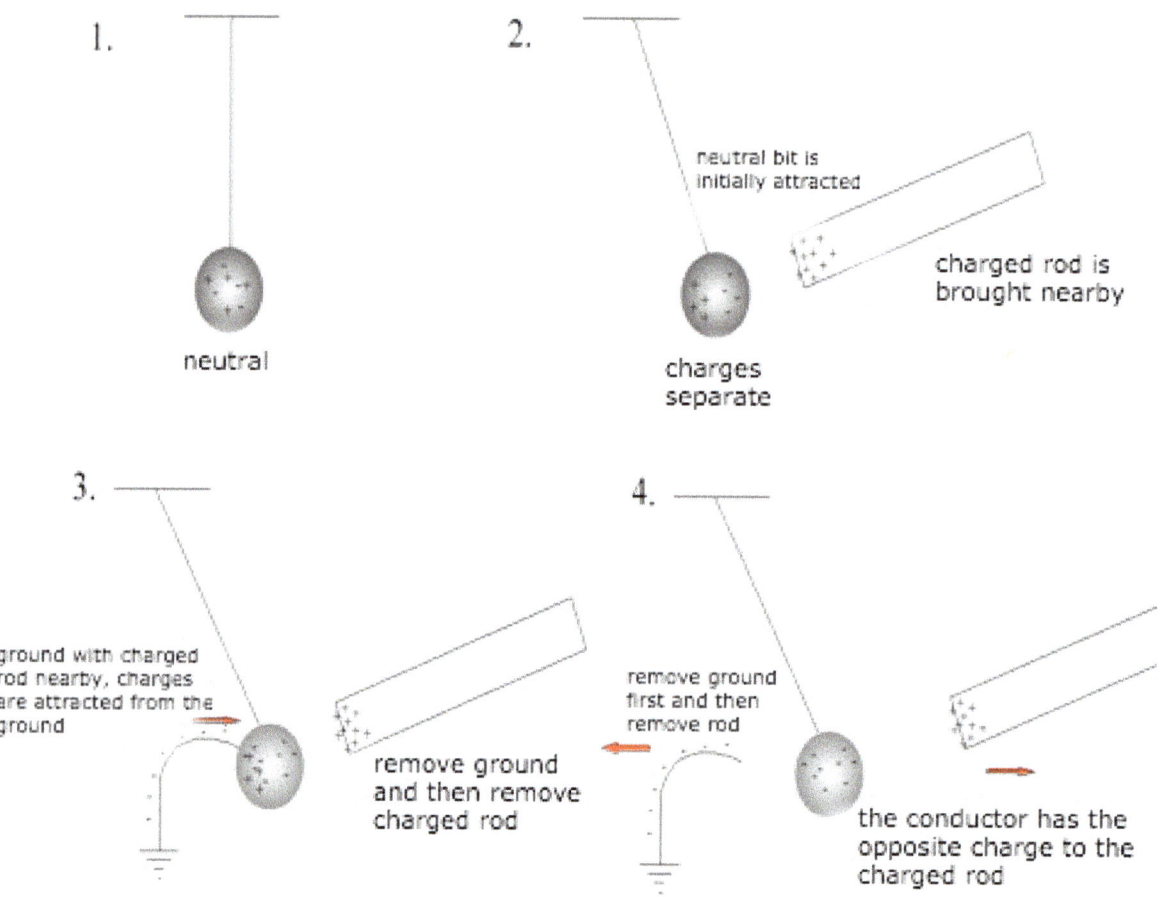

TN: Now is a good time to review some vocabulary that you have been using in your demonstrations. Using the correct terminology and understanding the movement of charge is important. Students will have an opportunity to complete their own activities and diagrams in the following exercises. Remember, as shown in the diagrams above, every time there is a a change from nearby to touching, a new diagram is required.

Electrostatic Lab Activities

Learning Activity 4.1: Attraction of Neutral Objects

TN: Students have now developed a particle model of electricity and they have some experience with the teacher demonstrations in explaining electrostatic phenomena. In the following hands-on exercises there are a number of interesting and fun electrostatic phenomena to investigate.

Materials: Plastic rods, pvc rods, paper bits.

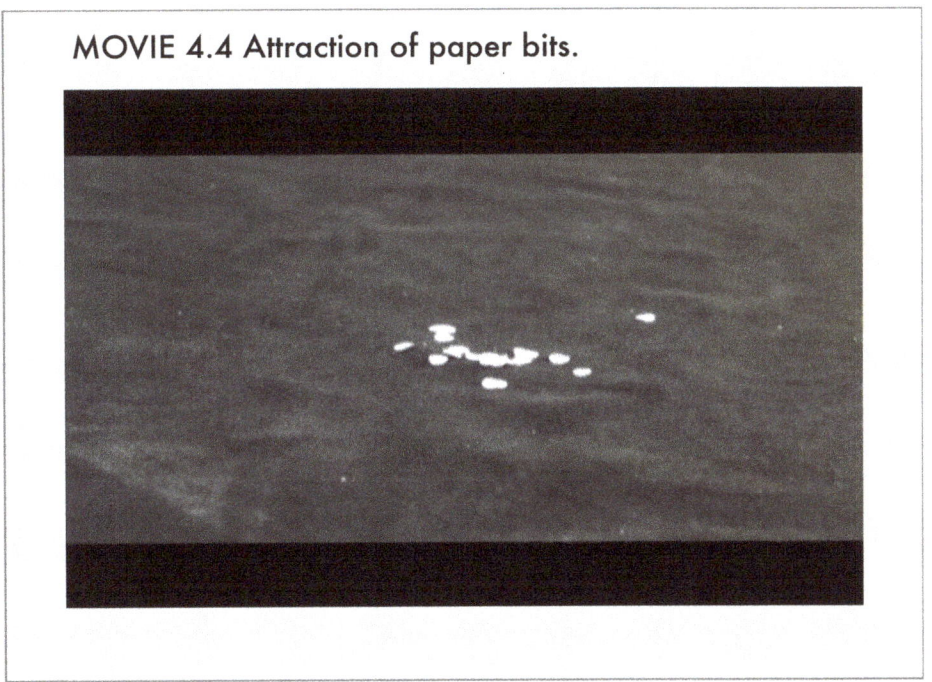

MOVIE 4.4 Attraction of paper bits.

https://www.youtube.com/watch?v=uYC3cFuToro

Instructions for the Activity

Perform the following activities, DIAGRAM YOUR EXPLANATION (every time a change occurs from touching to nearby or nearby to touching make a new diagram). Use the model for electric charge showing the movement of the negative charges, the net charge and the resulting actions. NOTE: We are calling the plastic rod rubbed with silk positive and the PVC pipe rubbed with wool negative (the distinction is arbitrary at this point).

Questions

<u>4.1a.</u> What do you predict will happen if you bring charged objects nearby each other?

TN: They will attract or repel.

<u>4.1b.</u> Take a few paper punch-outs and bring them nearby each other. What do they do to each other? What charge is on the punch-outs?

TN: Nothing. They are neutral.

<u>4.1c.</u> Rub a plastic rod with silk to put a positive charge on the rod and bring it nearby the punch-outs, what happens? Diagram showing the charge.

Attraction of paper bits diagram

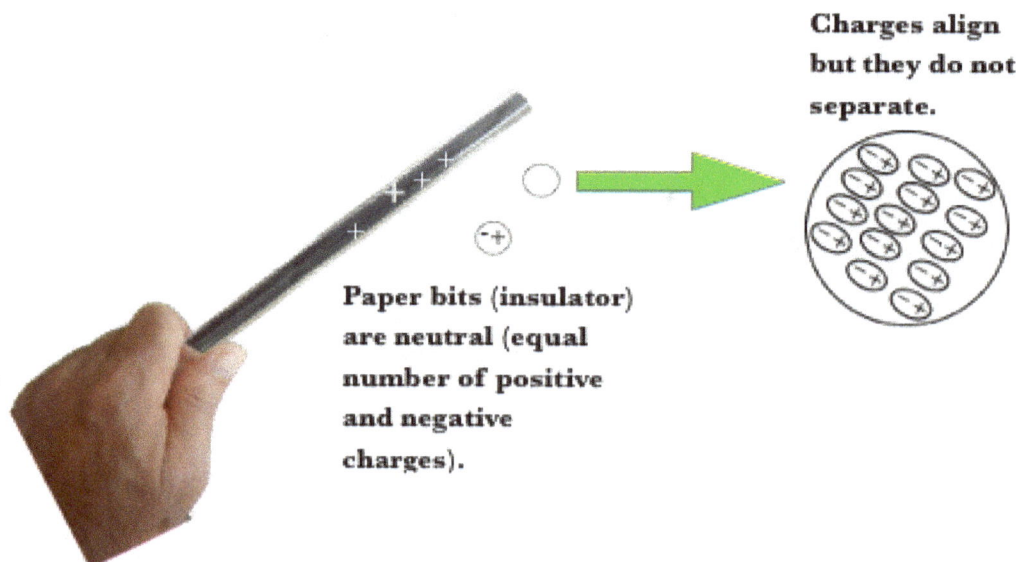

TN: The paper bits are attracted to the positive rod (charge pairs align in the paper bits with negatives closer).

<u>4.1d:</u>   Rub a PVC pipe with wool to put a negative charge on the PVC and bring it nearby the punch-outs, what happens?   Diagram showing the charge.   What do you conclude?

TN: The paper bits are also attracted to the negative rod. If the rod is negative the negative charges will rotate AWAY from the positive rod. There will still be a net force of attraction. Neutral objects are attracted to both positive and negative charges. Remember when charges are nearby charge will move in conductors and align (polarize) in insulators.

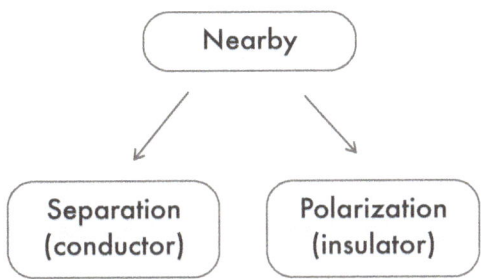

Learning Activity 4.2: Build an Electroscope

Materials: aluminum foil, nylon thread, styrofoam cup, plastic straw

TN: Students will need an electroscope for many of the following activities.

Instructions for the Activity

1. Tear off a small piece of aluminum foil about the size of a dime. Roll the foil between your fingers around a nylon thread to make form a small bit.

2. Cut a small notch in a plastic straw and suspend the nylon thread with the foil bit in the notch.

3. Punch a hole in a styrofoam cup and push the straw into the cup as shown. You can stack the cups to make your electroscope taller.

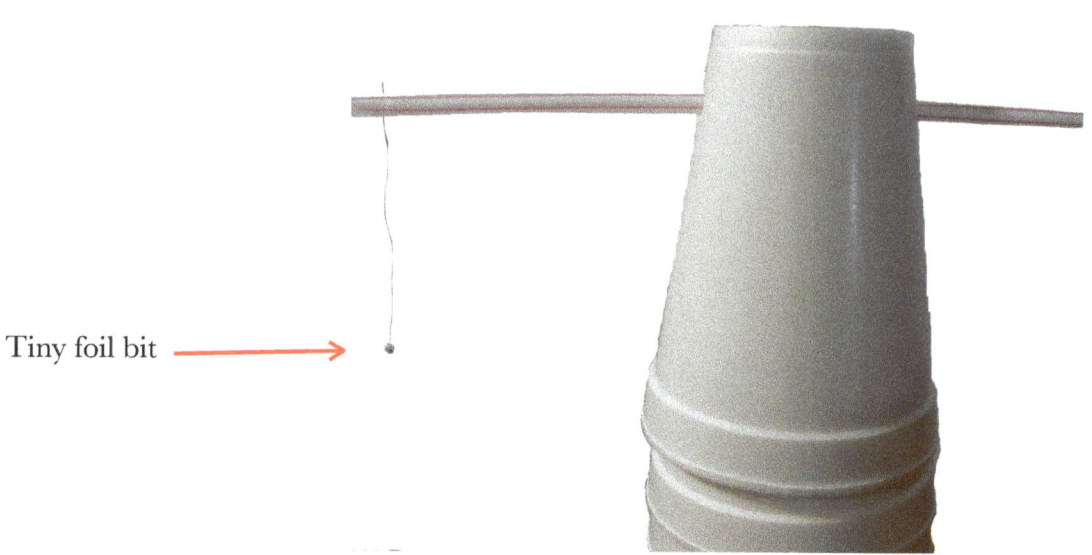

Tiny foil bit ⟶

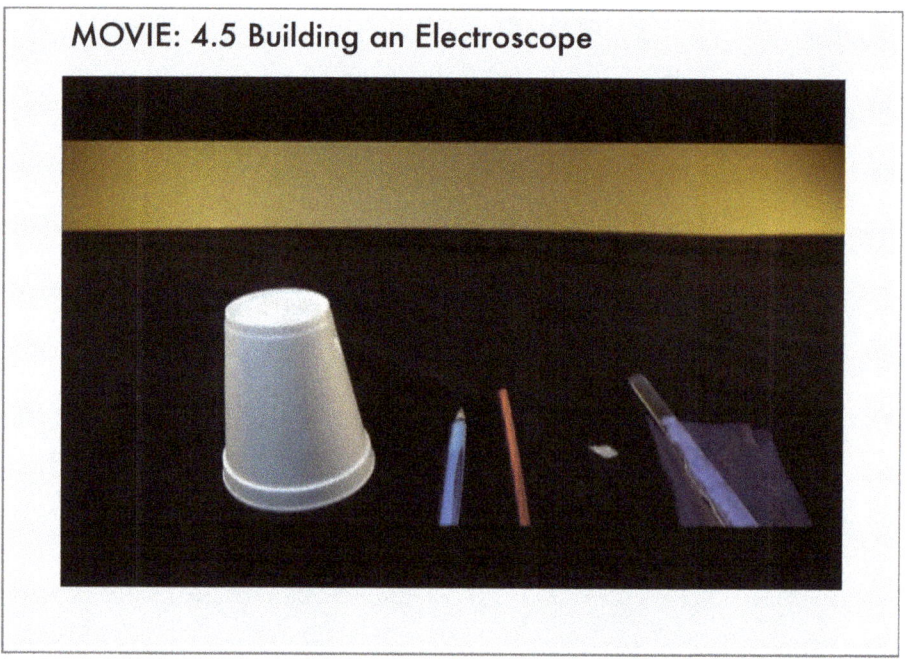

https://youtu.be/gBEOHbnzz-8

Essential Information 4.5: Testing for Charge

We use the electroscope to test for charge. In order to test an object for charge we put a known charge on the electroscope and bring unknown charges nearby. If they repel they are the same charge. If they attract they are either the opposite charge or neutral. Generally we can tell the difference between oppositely charged and neutral by the strength of the attraction but this is not definitive. The only true test for charge is one of repulsion. It is useful to have two electroscopes handy, one can be charged positive and one can be charged negative. However, if you only have one electroscope you can change the charge by neutralizing the foil bit by touching it and contacting with another charged rod.

Learning Activity 4.3: Testing for Charge

Materials: plastic rod, silky material, PVC rod, wool-like material, electroscope

Instructions for the Activity

1. Rub a plastic rod with silk to put a positive charge on the rod. Touch a suspended foil bit with the charged plastic rod. What charge does the foil bit have? How do you know?

TN: The foil bit is positive. Objects charged by contact have the same charge. The foil bit and charged rod repel each other.

2. Rub a PVC rod with wool to put a negative charge on the rod. Bring the charged PVC rod near the foil bit charged in step 1, what charge is on the PVC? How can you be sure?

TN: The foil bit is attracted to the PVC rod. It must be either negative or perhaps neutral. We can be sure by bringing the PVC rod nearby a negatively charged electroscope to see if it repels.

3. Bring the negative rod (PVC) near (but not touching) a neutral foil bit. Bring a positive (plastic) rod near the foil bit. Diagram each case. Why is the force of repulsion the only true test for charge?

TN: The neutral foil bit is attracted to both the positive and negative rods. The charges align as previously illustrated. The force of repulsion is the only true test for charge as neutral objects are attracted to both positive and negative charges.

Learning Activity 4.4: Attraction of a Neutral Insulator

Materials: plastic or PVC rod, ruler, bottle cap

Instructions for the Activity

1. Balance a wooden ruler on a plastic bottle cap and bring a charged rod nearby one end of the ruler. Explain why the ruler is attracted to the rod.

Turn the cap upside down.

TN: The ruler is an insulator. The positive negative pairs will align and result in a net force of attraction.

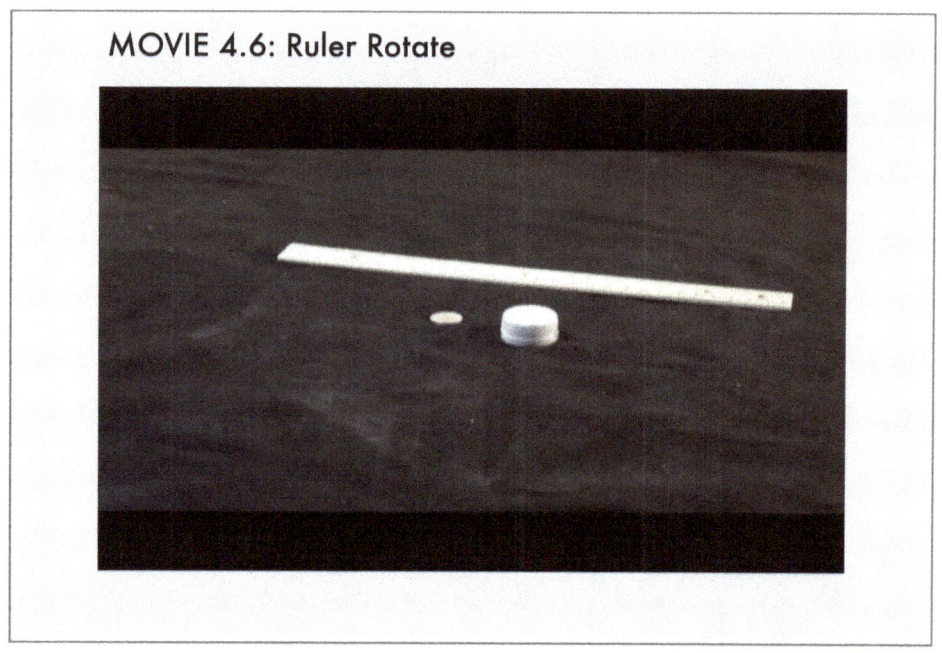

MOVIE 4.6: Ruler Rotate

https://youtu.be/uUGetq50iWE

Learning Activity 4.5: Conduction of Charge

Materials: charged rod, copper pipe, electroscope, styrofoam cup

Instructions for the Activity

1. Place a metal pipe on a styrofoam cup. Suspend a foil bit such that it touches one end of the pipe. Touch the other end of the pipe with a negatively charged rod.

Conduction of charge diagrams

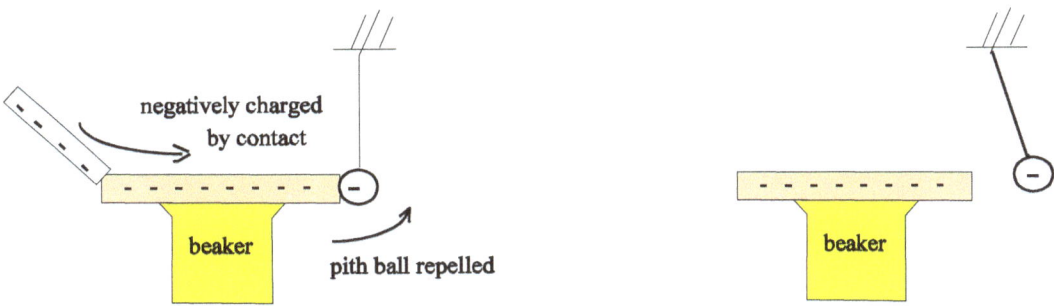

TN: The pipe is charged by contact, since it is a conductor the charge spreads out evenly across the copper. The foil bit is touching the copper so it picks up some charge and is repelled. If you use a positive rod the negative charges will flow onto the rod and the foil bit and copper pipe will be positive. They will repel.

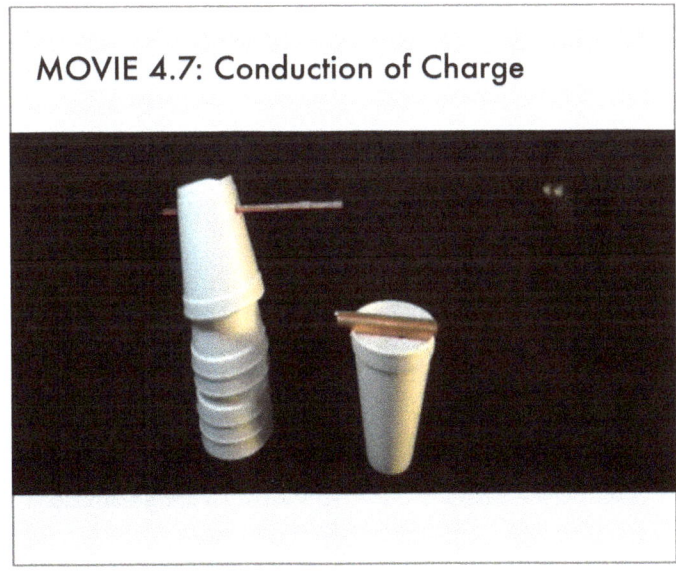

MOVIE 4.7: Conduction of Charge

https://www.youtube.com/watch?v=uYC3cFuToro

Learning Activity 4.6: Charging by Induction

Materials: charged rod, copper pipes, electroscope, styrofoam cup

Instructions for the Activity

1. Bring a charged rod close to one end of a single pipe and touch the other end of the pipe briefly with your finger.

2. Remove your finger with the rod nearby, then remove the rod and test the pipe for charge. Diagram the movement of charge.

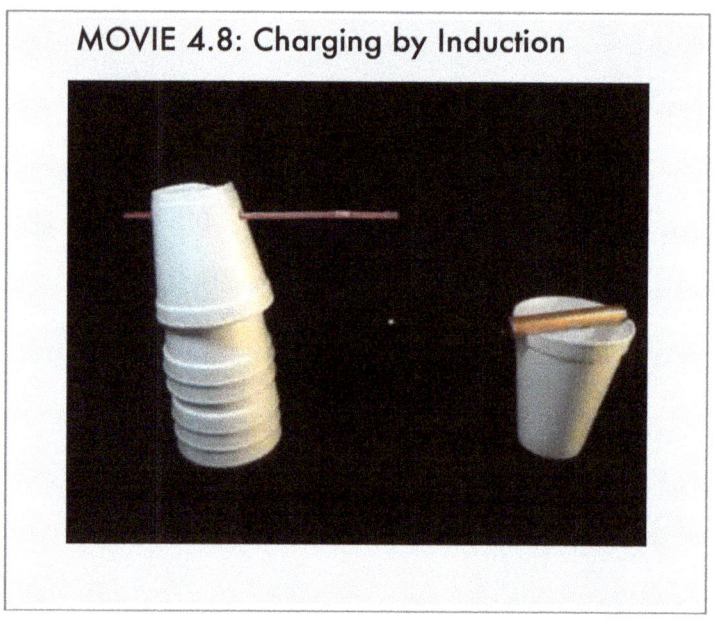

MOVIE 4.8: Charging by Induction

https://youtu.be/A9xpirGWXic

TN: Charging by Induction (1 pipe) diagram

1. If you bring a positive rod nearby, some of the negative charges will move towards the positive rod.. As a result, the far end becomes more positive. Remember we are showing the approximate net results - there are always lots of charges everywhere.

2. Ground the pipe by touching it at the far end. NOTE: You should feel a slight spark jumping from your finger to the pipe as negative charges in your finger are attracted onto the pipe.

3. Keeping the rod nearby, remove the ground (finger). Then, you can move the charged rod away. You have now "trapped" extra negatives on the pipe.

4. When the positive rod is moved away, the negative charges spread throughout the pipe and the pipe becomes negatively charged. Bring the rod nearby a negatively charged electroscope to test for charge.

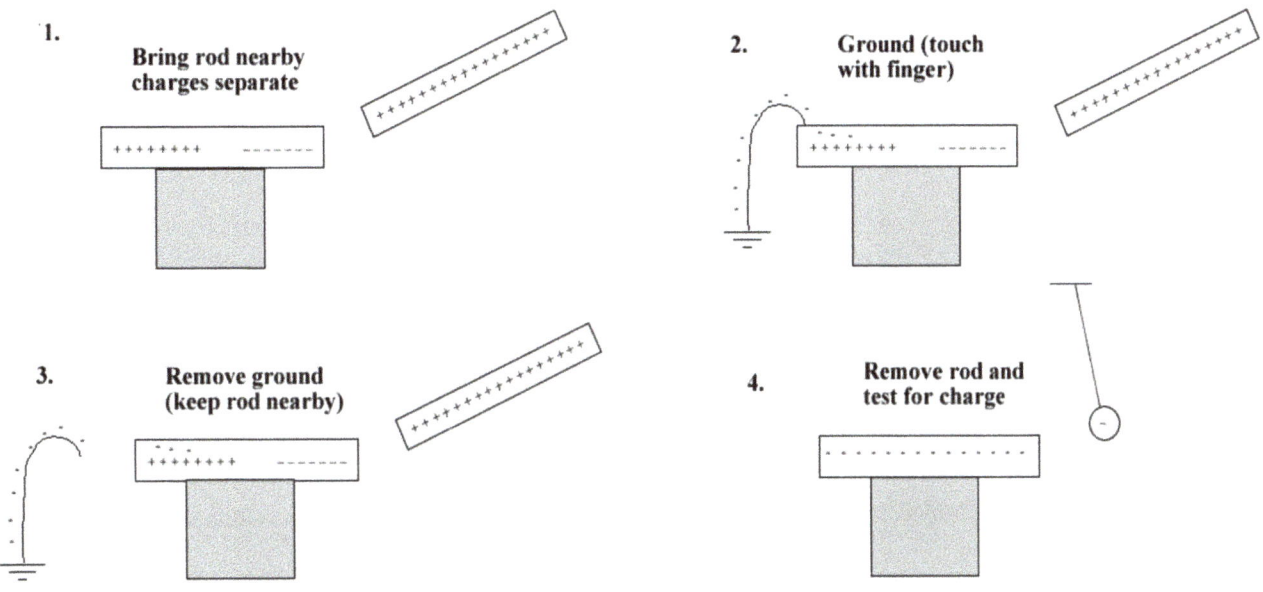

TN: Test the pipe for charge by bringing it nearby a charged electroscope (charge the electroscope first before you begin). The pipe will have the opposite charge to the charged object brought nearby.

NOTE: You can charge the pipe positively using a negative rod.

3. Place two metal pipes end to end so they touch each other and bring a charged rod nearby one end of a pipe (do not get the rod so close that a spark jumps). With the charged rod nearby, separate the pipes and test each pipe for charge. Diagram the movement of charge.

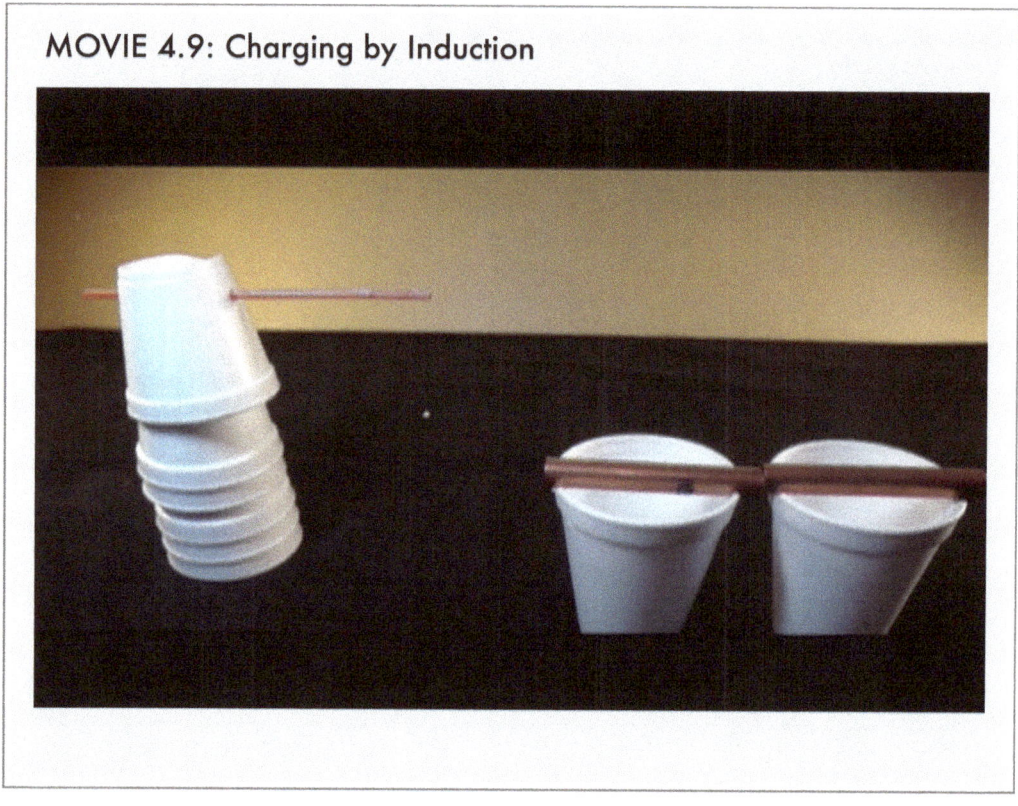

MOVIE 4.9: Charging by Induction

https://www.youtube.com/watch?v=MyOFB-QD9dw

TN: Charging by Induction (2 pipes) diagram

1. When the pipes are touching, bring a positive rod nearby one end. Some negative charges with be attracted and move towards the rod. The pipe on the right hand side gains negatives leaving the pipe on the left hand side more positive.

2. Keeping the charged rod nearby, separate the pipes (move the pipe on the left). When the pipes are separated you can move the charged rod away.

3. Test the pipe on the left by bringing it nearby a positively charged electroscope and the pipe on the right by bringing nearby a negatively charged electroscope.

Diagram

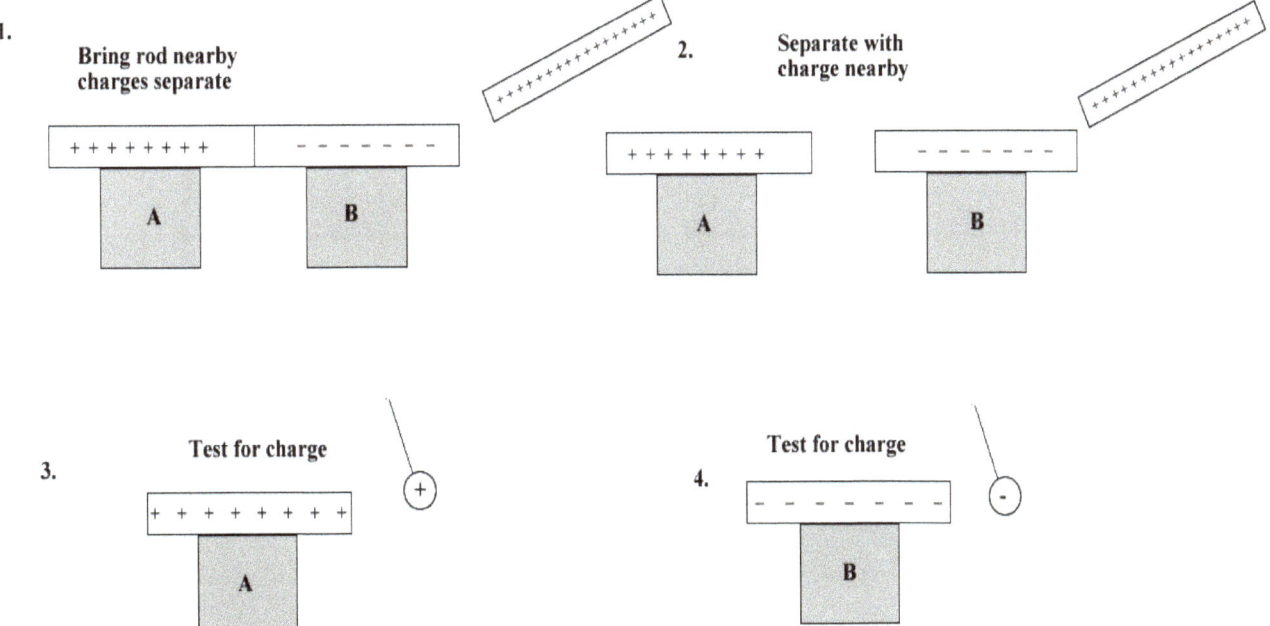

NOTE: You can reverse the charge on the pipes by using a negative rod.

Learning Activity 4.7: Inductive Force

Materials: charged rod, aluminum foil, ruler, round cup

Instructions for the Activity

1. Fold a piece of aluminum foil around a ruler into a long narrow strip (30 cm) like this:

2. Wrap the strip around a cup to form a loop and secure with some tape to make a wheel.

3. Place the wheel on a flat surface and bring a charged rod nearby and make your wheel roll! Diagram the electrostatic effects.

https://youtu.be/BmK8y-m17FY

Diagram

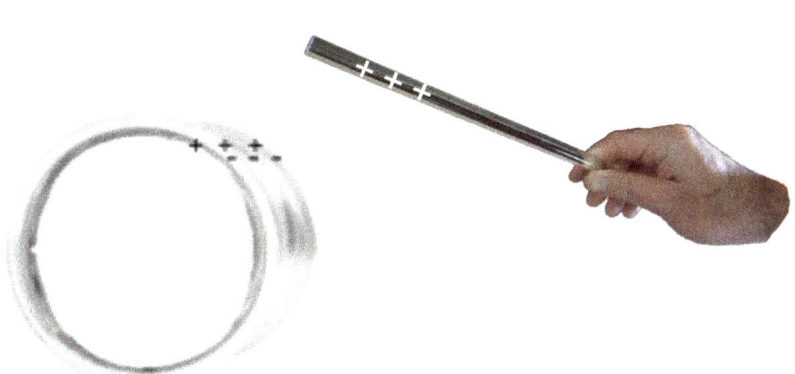

TN: The positive rod attracts negative charges nearby causing a net force of attraction. By positioning the rod and moving it you can make the wheel roll. Set up a race with one student on one side of a table and another on the other side. First person to roll their wheel over the end of the table wins. Tournament time!

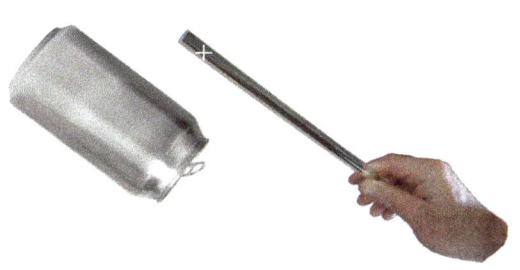

NOTE: You can try the same activity with an aluminum pop can.

TN: I like to use the student made foil wheel since the design plays a role in the student's success. Their wheels will vary depending on weight, circular integrity, and attracting technique.

Learning Activity 4.8: Oscillating Paper Bits

Materials: charged rod, paper bits, table

Instructions for the Activity

1. Bring a charged rod nearby a small number of punch-outs. Try to make the punch-outs oscillate.

TN: Even though a paper bit is an insulator, some charge may move.

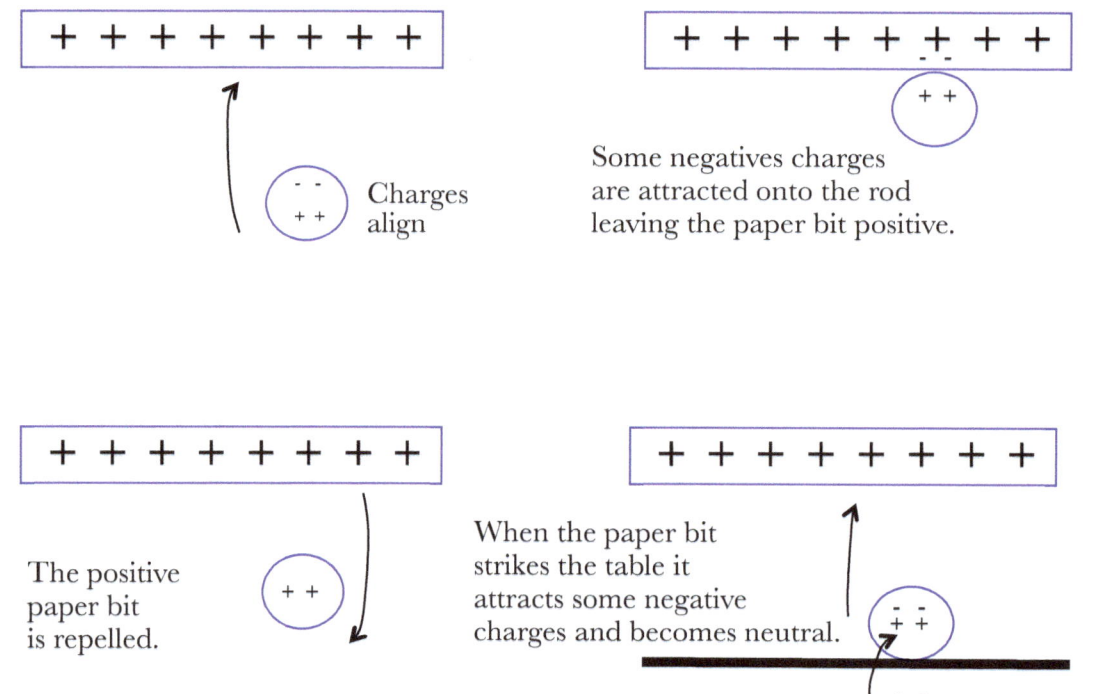

Learning Activity 4.9: Oscillating Foil Bit

Materials: charged pipes, styrofoam cups, electroscope

Instructions for the Activity

1. Suspend a foil bit between two pipes which are separated by about 10 cm (you can adjust this distance for maximum effect). Charge one pipe positive and the other negative. Can you make the foil bit oscillate?

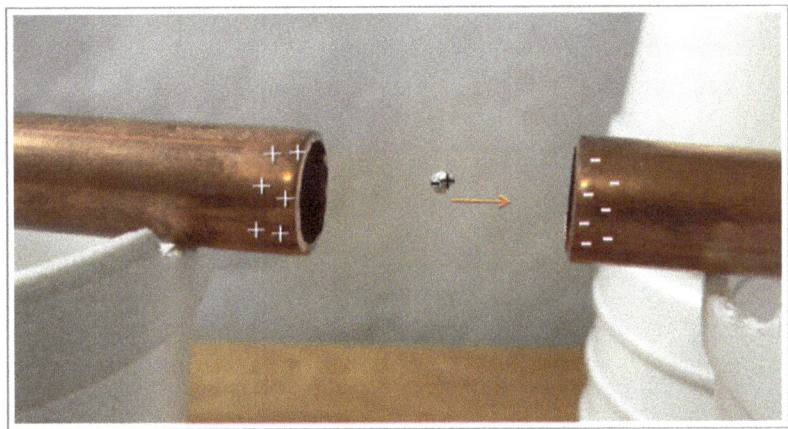

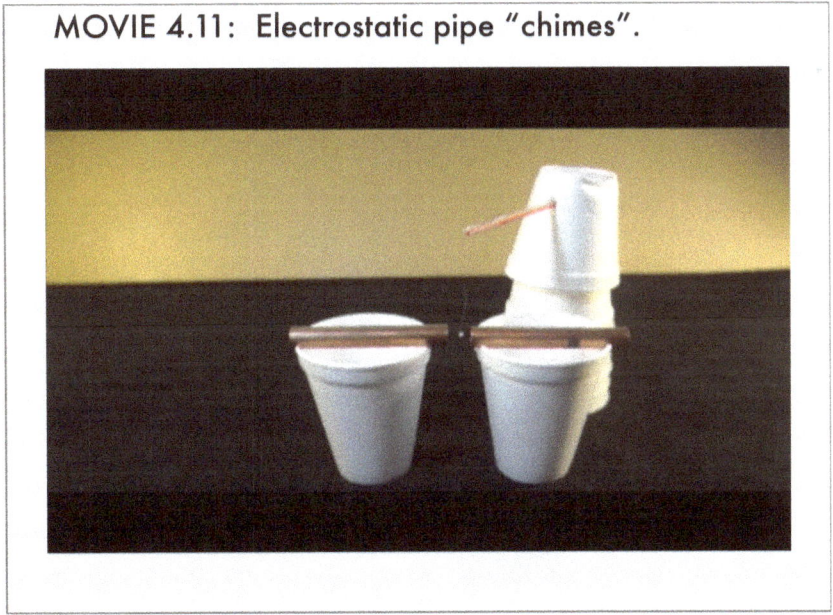

MOVIE 4.11: Electrostatic pipe "chimes".

https://youtu.be/CpazGlOxwZQ

TN: The foil bit is initially neutral so it is attracted to one of the charged pipes - in the example it is the negative pipe (1). When the foil bit touches the pipe it acquires some negative charges (2) and it is repelled by the negative pipe and attracted by the positive pipe (3). Then, it touches the positive pipe and gives up some of its negative charge and becomes positively charged (4). The foil bit is again repelled and attracted to the other side (5) and the cycle repeats. This can happen very rapidly and will continue until the negative charge on one pipe is transferred to the other. Students should diagram each change showing the movement of charges. Remember to ask where do the negatives go?, from where to where?.

Diagram

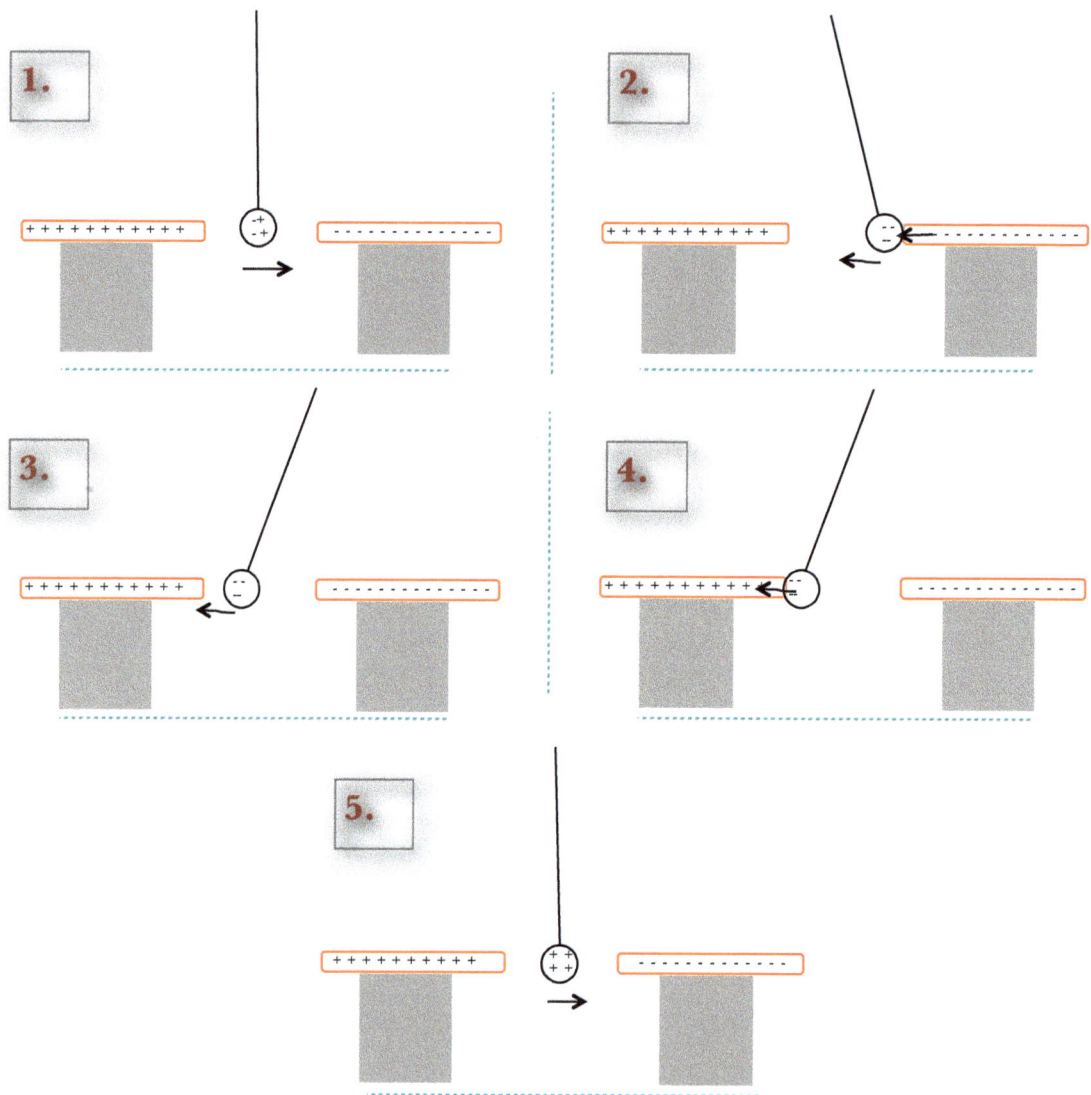

CHAPTER 5

Volta's Electrophorus

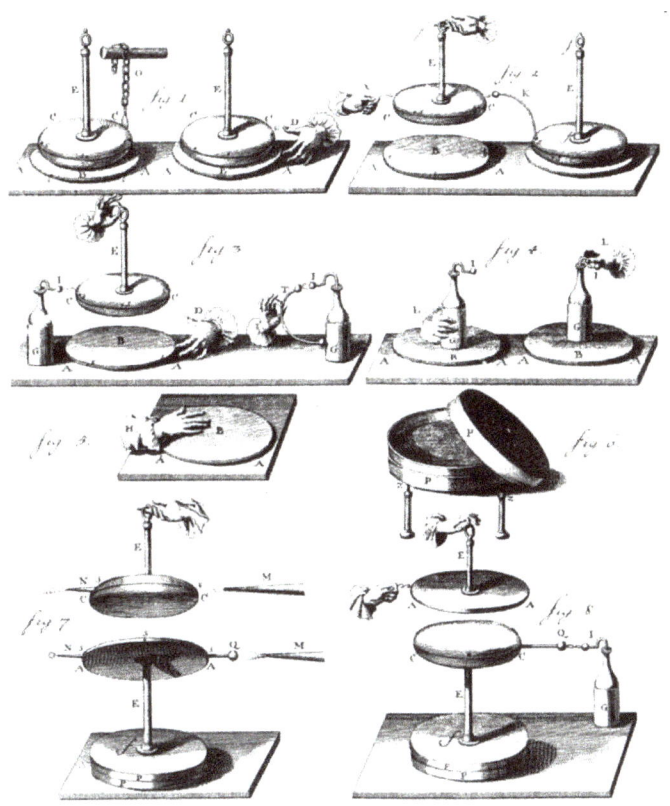

Introduction

As early scientists began to investigate electricity they tried to invent devices that could generate and store electric charge. The Italian scientist Alessandro Volta improved upon an earlier invention of the electrophorus and demonstrated how it could be used to study electrostatic phenomena.

Objectives:

1. Investigate and explain electrostatic phenomena using the particle model of electricity. Include conservation of charge, conduction, grounding, attraction of a neutral insulator, and induction.

Learning Activity 5.1: Volta's Electrophorus

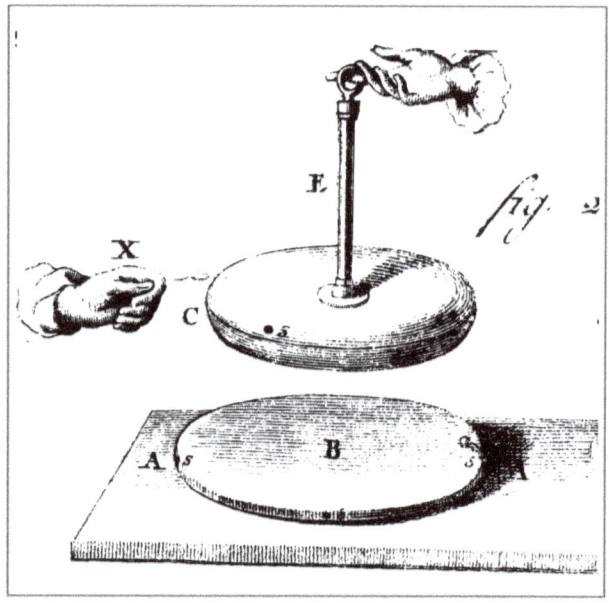

TN: Volta's electrophorus is a device which can be constructed by the students with very simple materials. This electrophorus is used to demonstrate charging, grounding, induction, conduction, and transfer of charge. If a student is able to explain all of the phenomena they observe with their electrophorus they will have a very good understanding of electrostatics. You can find the materials to build the electrophorus at local stores. Students love this activity.

Materials: Stryofoam plate, styrofoam cup, aluminum pan (pieplate), silk thread (fine nylon fishing line will do), foil piece, tape, straw.

Instructions for the Activity

1. Build your electrophorous as shown in the diagram. The foil bit should hang straight and touch the side of the pieplate.

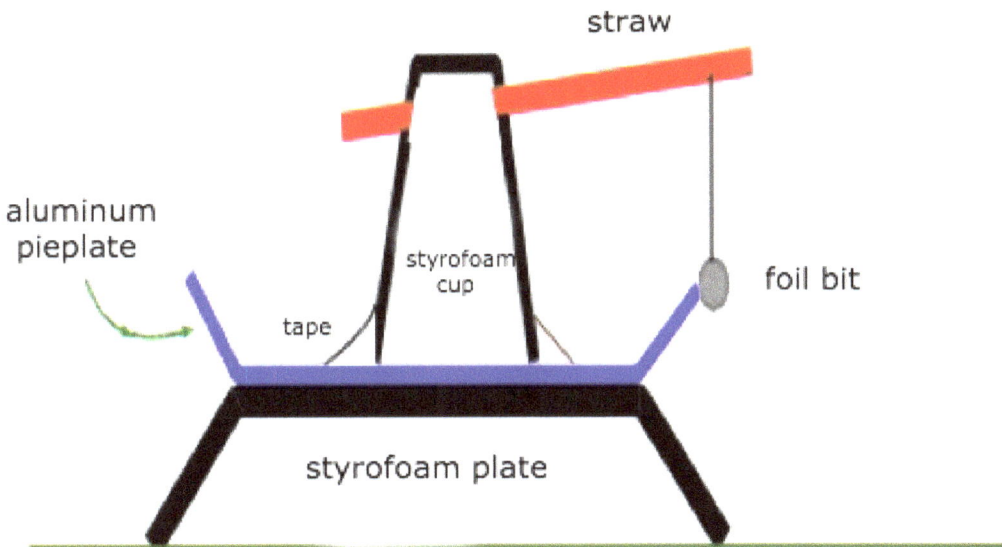

TN: The best way to approach this activity is to let students explore and answer the questions on their own or in a group of two. Groups of three students with two electrophorus plates also works well. After a few minutes the teacher should review the questions with students and complete corresponding explanations with diagrams. Finally, challenge the students to use their electrophorus to light the LED bulb continuously as a way to transition from static electricity to current electricity (book 2). The teacher should repeatedly ask the students, "Where do the negative charges go?", "From where to where?", "What is the result?".

2. Rub the styrofoam plate with wool and test for charge. What charge is on the styrofoam? How can we be sure?

TN: The charge on the Styrofoam plate will be negative. We can be sure by bringing the plate nearby a negatively charged electroscope – the plate and the electroscope will repel.

MOVIE 5.1: Charge on a pieplate.

https://youtu.be/jU-WOOGkgLE

2. Place the aluminum pieplate on top of the charged styrofoam. What happens to the foil bit? What charge do you think is on the pieplate? Diagram and explain.

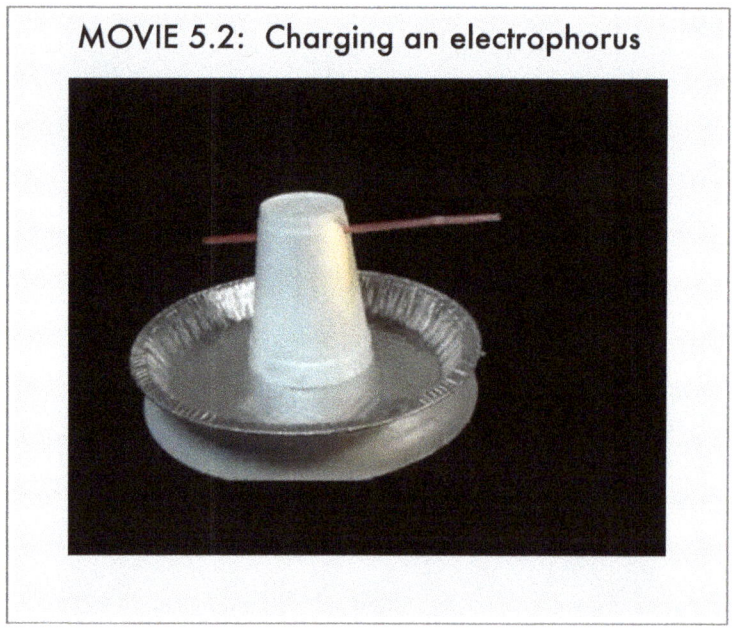

MOVIE 5.2: Charging an electrophorus

https://youtu.be/Fp4Vzau0qaA

Diagram: The Electrophorus

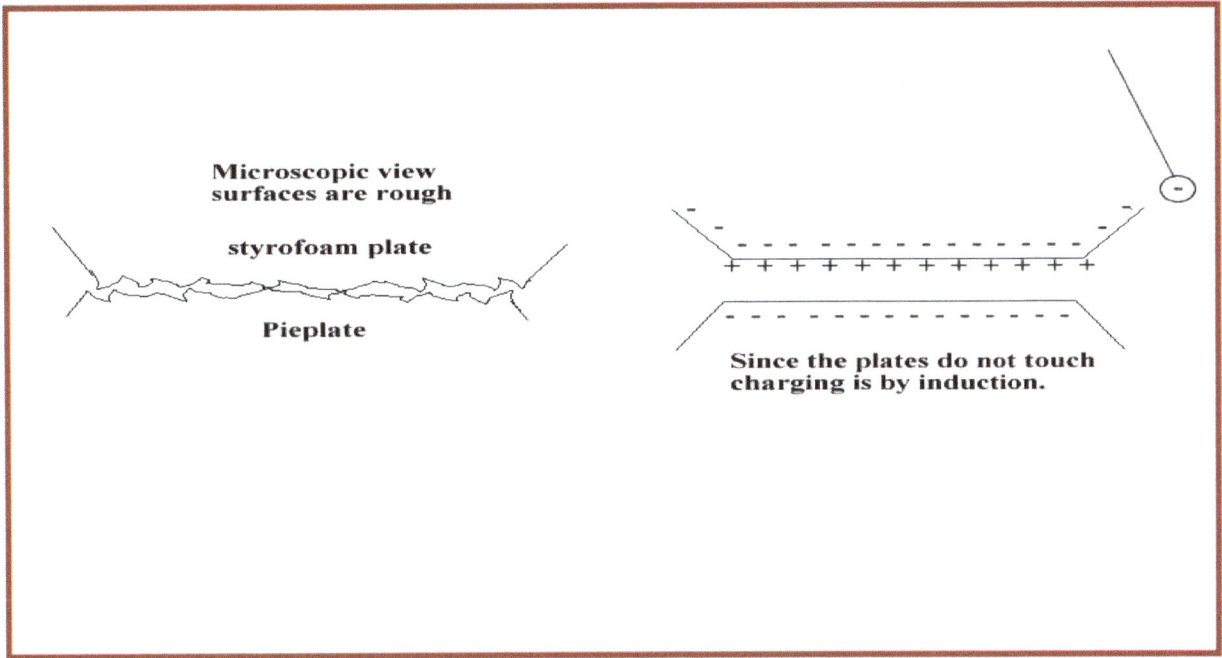

TN: The foil bit will repel (Note: be sure that the foil bit is touching the plate to begin). Students will typically answer that the styrofoam plate is negative as charges were transferred from the styrofoam to the aluminum plate (NOTE: THIS IS **NOT** CORRECT).

Essential Information 5.1: Charging by Induction

Microscopically the two surfaces only touch in a few points and since the styrofoam is an insulator charge does NOT transfer to the pieplate. However, since the charges on the styrofoam are nearby they cause the charges on the pieplate to separate. Negative charges on the pie plate move further away and some transfer to the foil bit. The foil bit is repelled from the side of the plate. The electrophorus is charged by induction.

Learning Activity 5.2: Charging an Electrophorus by Induction

Materials: Pieplate electroscope

Instructions for the Activity

1. Lift the pieplate off of the styrofoam plate by touching ONLY the insulating cup. You can hold the styrofoam plate down with your free hand. What happens to the foil bit? What charge is on the pieplate? How do you know?

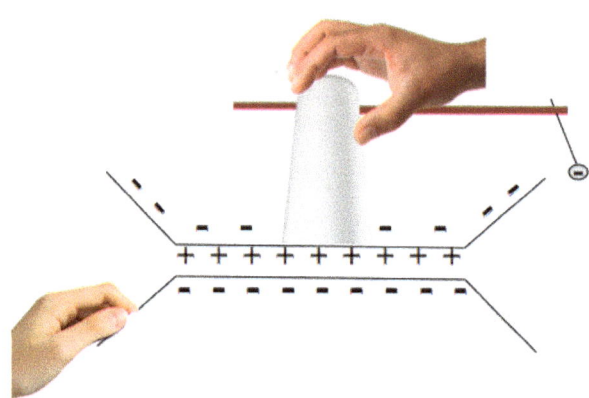

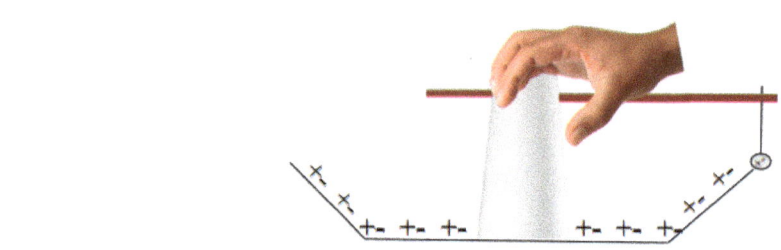

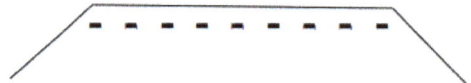

TN: When you lift the pieplate off of the charged styrofoam plate, the foil bit swings back and touches the side of the aluminum plate again. The aluminum plate must be neutral, as the foil bit is not being repelled.

NOTE: This is a discrepant event for students – why does the plate appear to be charged when it rests on the styrofoam and why is it neutral when it is lifted off of the styrofoam? When the pieplate is removed from the styrofoam the charges distribute evenly and all parts of the plate are neutral.

2. Place the pieplate back on the styrofoam. Be sure the foil bit repels from the side of the pieplate. If not, lift the pieplate and touch it, then start over by rubbing the styrofoam. With the pieplate resting on the styrofoam and the foil bit repelling from the side of the plate, touch the pieplate with your finger. What do you feel? What happens to the foil bit?

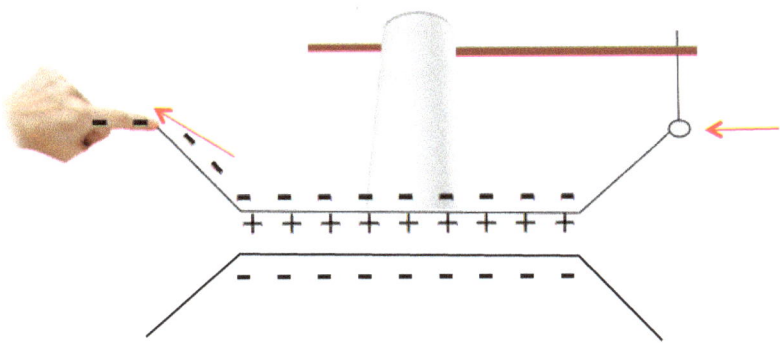

Diagram

TN: When the foil bit is repelled from the side of the pieplate and you touch the pieplate you will feel a mild shock. You can hear, and in the dark, see a small spark jump from the plate to you finger (NOTE: the shocks are very small like the shock you may feel after walking over a carpet and touching a door knob). The foil bit swings back and touches the pieplate. The negative charges on the pieplate are repelled by the negative charges on the styrofoam (remember the plates are not really touching!) and follow a path to ground through your finger.

3. Lift the pieplate off of the styrofoam by touching <u>ONLY</u> the insulating cup and holding the styrofoam plate down if needed. What do you feel? What happens to the foil bit as you remove the pieplate? What charge do you think is on the pieplate? Diagram and explain using the model for electric charge. Test your predictions.

Diagram

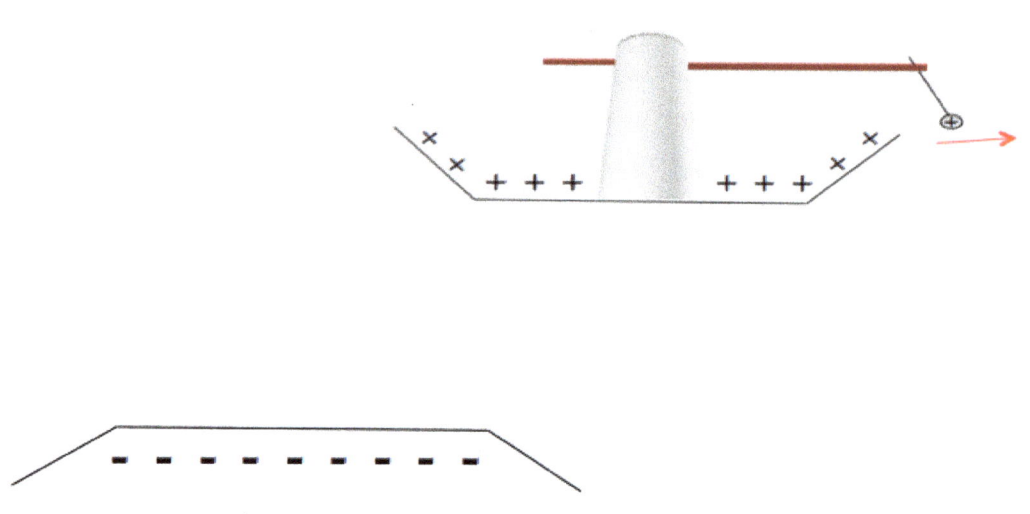

TN: As you try to remove the pieplate you should feel a small force holding the two plates together. You can hold down the styrofoam plate. When the plates are separated the foil bit repels (indicating that the pieplate is now charged). Since negative charges were removed the pieplate has excess positive charges. The charge is distributed evenly and the foil bit repels from the side of the plate. The charge on the pieplate is positive, you can test for charge by bringing nearby a positively charged electroscope.

TN: The plate was charged by induction. You can touch the plate to remove the charge (ground it) and repeat the process several times. The diagram is the same as the diagram for charging by grounding. The interesting part about charging in this manner is that no charge on the styrofoam is ever transferred and the process may be repeated over and over (However, eventually the charge on the styrofoam will dissipate into the air).

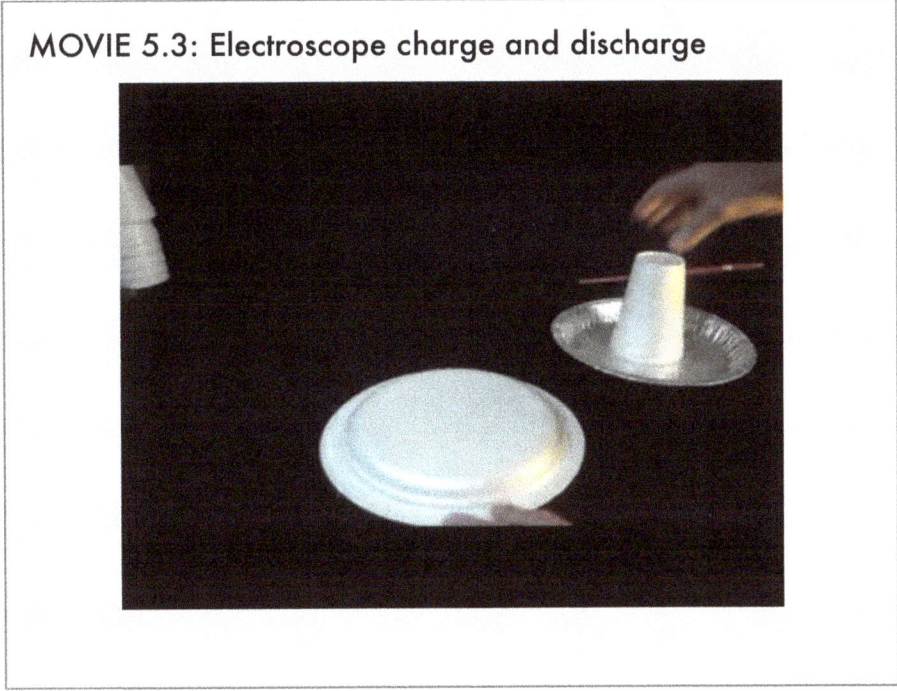

MOVIE 5.3: Electroscope charge and discharge

https://youtu.be/nfx9X-sEG_U

Learning Activity 5.3: Oscillating Foil Bit

Materials: Pieplate electroscope

Instructions for the Activity

1. Charge the pieplate as described previously (the foil bit should be repelled from the side of the plate. Bring your finger nearby and touch the foil bit. Explain your observations using the model for electric charge.

TN: As you bring your finger nearby the foil bit it will oscillate back and forth. The positively charged foil bit is attracted to the neutral finger (2), when they touch negative charges from the finger are transferred to the foil bit making it neutral (3). Then, the neutral foil bit is attracted back to the positively charged pieplate (4). When the foil bit touches the pieplate negative charges on the foil bit are attracted onto the pieplate (5). The pieplate becomes a little less positive and the foil bit becomes a little bit more positive. The positively charged foil bit is now attracted to the neutral finger and the process repeats – at times very rapidly (6). Another fun activity – be sure to continually ask them, where do the negative charges go, from where to where?

Diagram for an oscillating foil bit

1.

2.
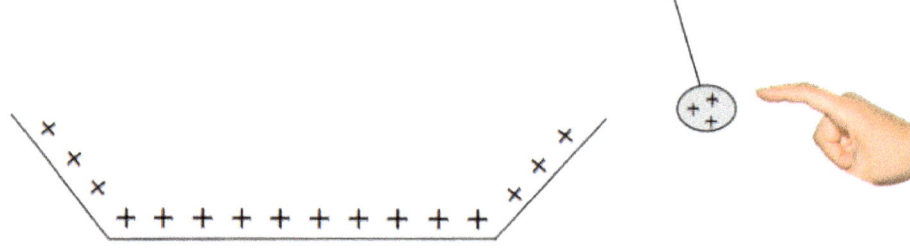

3.
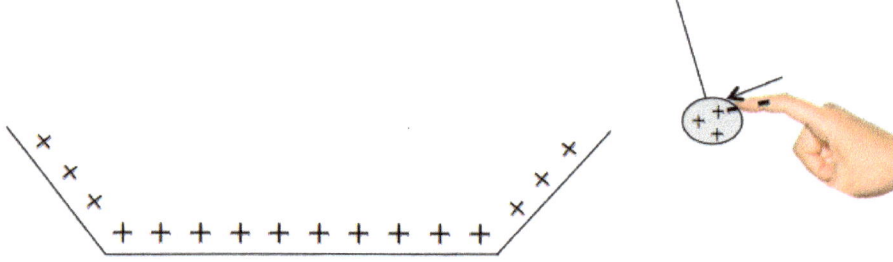

4.
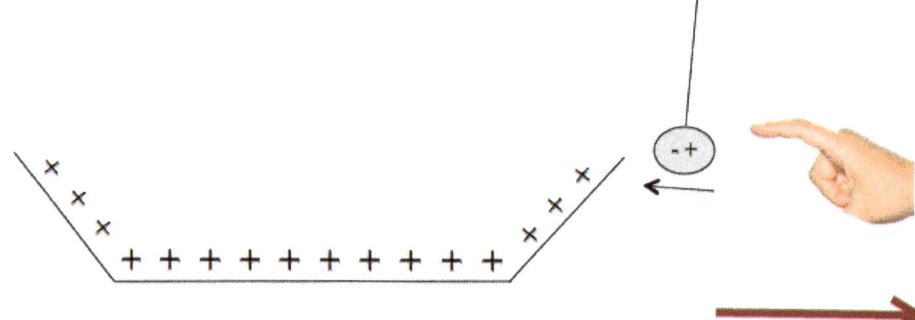

5.

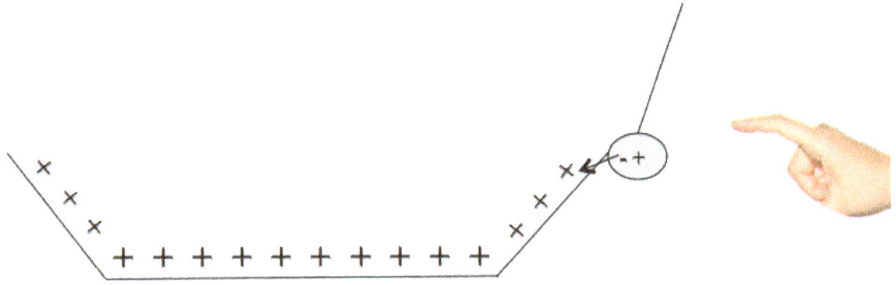

6.

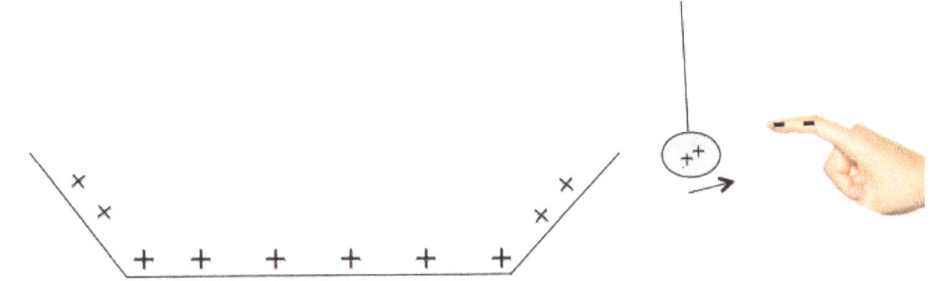

MOVIE 5.4: Oscillating foil bit.

https://youtu.be/ETfAHDecRA0

Learning Activity 5.4: Lighting an LED bulb

Materials: Pieplate electrophorus, LED bulb (type NE 2).

Instructions for the Activity

1. Charge the pieplate. Hold one end of a neon LED bulb between your fingers and touch the other end to the charged electrophorus. Observe the bulb. Diagram and explain using the model of electric charge.

TN: The pieplate will be charged positively. When one end of the neon bulb touches the pieplate, negative charges from ground (the finger) pass through the neon bulb onto the pieplate. As the negative charges pass through the neon gas in the bulb they collide with the particles in the gas and give off energy as light. The process stops when the plate becomes neutral – this happens very rapidly.

Diagram for transitional current

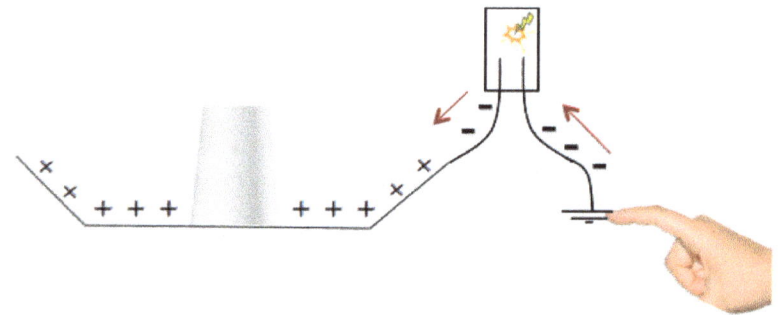

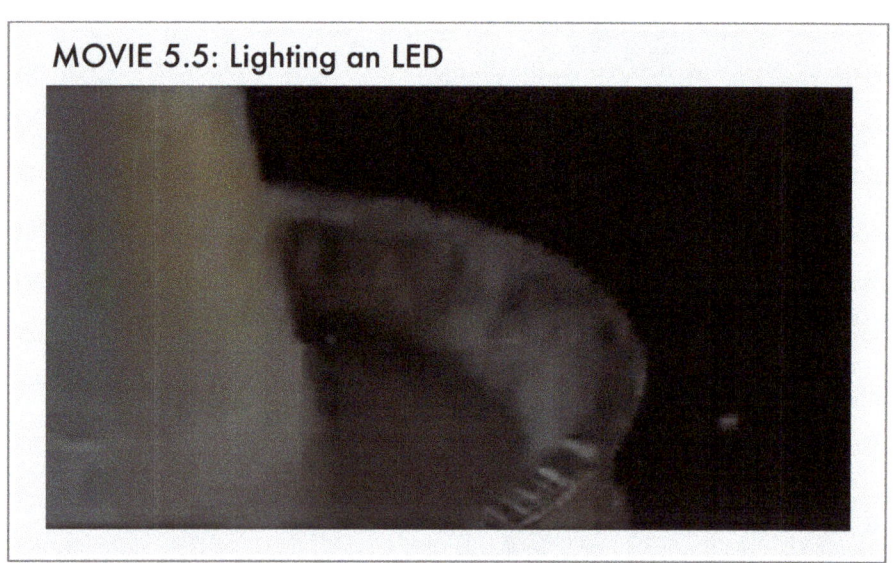

MOVIE 5.5: Lighting an LED

https://youtu.be/0ntcZoQOlaY4b.

Student Challenge: Can you keep the bulb lit? How? Demonstrate.

TN: There are many ways to try to keep the bulb lit. Essentially, since you very quickly run out of charge you must continually add charge to one end of the bulb. Students often try this by moving the pieplate up and down off of the styrofoam or by touching the side of the pieplate with the bulb as they continually rub the Styrofoam plate with the pieplate pushed off to one side. Let them experiment, there is no right answer.

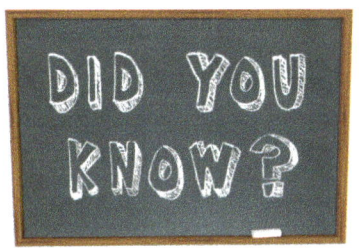

Understanding the electrophorus was difficult for many of the great scientists and Volta himself. The electrophorus quickly gained fame for Volta throughout Europe, not the least of the reasons being that Volta wrote many letters describing the device to important scientists and he sent many models as gifts to influential people. The electrophorus, however, was not without controversy or mystery. Some said that its seemingly endless supply of charge challenged the wits of the most celebrated scientists of electricity in Germany and Italy. In London, in a presentation on the electrophorus before the Royal Society, a local scientist lamented that "It is hard to say how or where the electricity is deposited; there is so much of it." Benjamin Franklin, on the other hand, when hearing about the device, immediately recognized it as a type of Leyden jar. Eventually, everyone came to understand that microscopically the plates did not touch and that the charge was induced on the metal plate.

In this famous portrait of Allesandra Volta, Volta is holding a book while in front of him on the table is his battery and an electrophorus. What do you think is the significance of these items?

TN: Volta had great pride in his apparatus but the explanation (book) was most important and is held closer to his heart.

Learning Activity 5.5: The Leyden Jar

Materials: film canister (or similar), aluminum foil, paper clip

Instructions for the Activity

1. Carefully coat the inside and outside of a film canister with aluminum foil.

2. Push a paper clip through the lid so that it touches the foil on the inside of the canister.

3. Charge the Leyden jar with a plastic rod rubbed with silk by touching the paper clip many times.

4. Hold the canister between your thumb and finger and touch the paper clip with a finger from the same hand. Explain the charging and discharging of the Leyden jar.

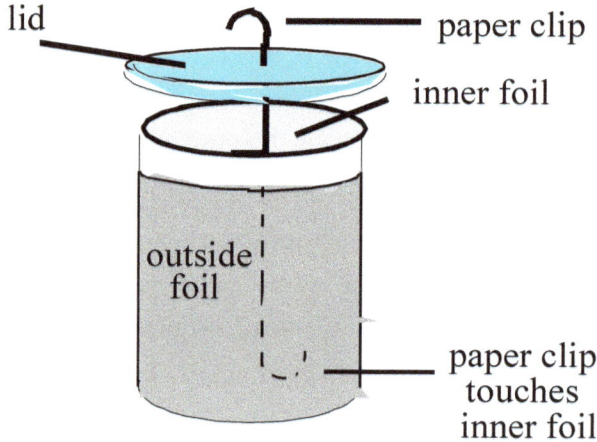

FILM CANNISTER LEYDEN JAR

TN: The Leyden jar is used to store charge. The charge placed on the inside of the Leyden jar induces the opposite charge on the outside of the jar. These opposite charges attract each other and hold the charge on the jar. This was the first storage device for charge. When you touch the outside and inside of the jar, charge flows and the Leyden jar is neutralized. Some commercial Leyden jars (often found in school labs), can hold a considerable amount of charge and should be handled **VERY** carefully if used. Our small example here also works very nicely with the LED bulb (touch one end to the paper clip and the other to the side of the Leyden jar.

Appendix: Equipment

Most of the materials used in the student activities are easily obtained at local stores. Plastic and PVC rods can be bought at your hardware store and cut to length. Special apparatus such as the neon LEDs (NE 2 type) can be purchased through a science supply vendor or by checking online sources such as Ebay.

Don Metz

If I had an hour to solve a problem and my life depended on the solution, I would spend the first 55 minutes determining the proper question to ask, for once I know the proper question, I could solve the problem in less than five minutes.

- Albert Einstein

About The Author

Dr. Don Metz was a Full Professor in the Faculty of Education at the University of Winnipeg, Canada until 2015 and is now a Senior Scholar. Don holds a BSc in Physics, and a MEd and PhD in Science Education. Don taught physics, mathematics and general science at the high school level for twenty years before completing his PhD and moving into the Faculty of Education at the University of Winnipeg. He has been extensively involved with curriculum development in the province of Manitoba as the principal writer of the Manitoba Physics program and as author of several teaching resources including IN MOTION and TEACHING ELECTROSTATICS. He has been very active in the professional development of in-service teachers for many years. Don has presented at numerous conferences and published papers in a wide variety of education journals including the Teacher Education Journal, Science Education, Science & Education, The Physics Teacher, The Mathematics Teacher and the Canadian Journal of Environmental Education.

www.ingramcontent.com/pod-product-compliance
Lightning Source LLC
Chambersburg PA
CBHW061145010526
44118CB00026B/2876